So You Want To Be
A Consultant

So You Want To Be

A Consultant

Harry E. Chandler

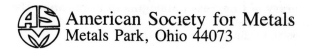 American Society for Metals
Metals Park, Ohio 44073

Library of Congress Cataloging in Publication Data
Chandler, Harry E., 1920–
   So you want to be a consultant.

   Bibliography: p.
   Includes index.
   1. Consulting engineers—Handbooks, manuals, etc.
I. Title.
TA216.C49   1984   620'.0068   84-6385
ISBN 0-87170-184-7

SAN 204-7586

To Robert L. Ray, members of the Independent Metallurgical Engineering Consultants of California, and the 30 other consultants who made this book possible.

Preface

In doing homework for *So You Want To Be a Consultant,* I read a dozen or more books and references. In time it occurred to me that each volume was limited to presenting the views and experiences of a single author, or at most two. What if I could put together an anthology of views and experiences built from the input of an impressive number of established consultants?

Time and travel required to interview a multitude of sources were a discouraging prospect, not to mention the practical aspects of identifying and locating these special people.

I eventually found a way to bypass the first problem by developing the equivalent of an in-depth personal interview on paper—a 17-page questionnaire containing 120 questions.

I needed and got outside help on the second problem from essentially two sources:

First, from Bob Ray, a metallurgical engineering consultant and friend based in Oakland, CA. Through him I secured the cooperation of IMECA (the Independent Metallurgical Engineering Consultants of California). Fifteen of its 20 members, including Bob, took part in the survey; they are listed at the end of this preface. They are more completely identified in the directory of consultants (Chapter 30). Their assistance is gratefully acknowledged.

Second, from the American Society for Metals. Thirty members of ASM who are consultants took part in the survey. They are named at the end of this preface, as well as in the directory of consultants. Their assistance is gratefully acknowledged.

Finally, to add a dimension to the counsel of the panel of consultants, I surveyed a small group of clients—employers of consultants—separately. This much shorter questionnaire was tailored to get the client's point of view on a variety of topics discussed by the consultants. Roughly 75 per cent of the members of this industry panel are presidents of companies. The remainder are vice presidents. Of the 11 respondents, two chose anonymity. The others are named at the end of this preface. Their assistance is gratefully acknowledged.

First, the IMECA group: F. Sidney Badger, C. Howard Craft, Bernard F. Faas, John A. Fellows, Robert G. Ford, Gerald P. Fritzke, Carl E. Hartbower, Abraham Hurlich, Augustus B. Kinzel, Larry

E. McKnight, George E. Moller, Robert E. Monroe, Austin Phillips, Robert L. Ray, Russell G. Sherman.

The ASM group: Clo E. Armantrout, Dr. William W. Austin, Dr. D.J. Blickwede, Roger W. Bolz, Nick F. Bratkovich, Dr. Robert M. Brick, Joseph E. Burke, Dr. Hallock C. Campbell, Joe Devis, Allen B. Dove, W. William Dyrkacs, Muir L. Frey, Bruce W. Gonser, Richard W. Hanzel, Earl T. Hayes, John P. Howe, J.R. Kattus, Walter A. Luce, Ray F. Kern, Norbert K. Koebel, Dr. Bernard S. Lement, Dr. N.E. Promisel, Bernard R. Queneau, Philip C. Rosenthal, James M. Shults, Dr. Harold Jack Snyder, R. David Thomas, Jr., Donald J. Wulpi, Dr. Carl A. Zapffe.

The client group: Wally L. Bamford, president, Can-Eng Sales Ltd., Niagara Falls, Ont.; Dr. Arden L. Bement, Jr., vice president, technical resources, TRW Inc., Cleveland; Beresford Clark, president, Metallurgical Processing Inc., Ft. Wayne, IN; Jess F. Helsel, president, Helsel Metallurgical Inc., Campbellsburg, IN; John D. Hubbard, president, Hinderliter Heat Treating Inc., Tulsa, OK; Frederick H. Perfect, president and research director, Reading Alloys Inc., Robesonia, PA; Dr. T.R. Pritchett, vice president, research & development, Kaiser Aluminum & Chemical Corp., Pleasanton, CA; John H. Ries, vice president of operations, Lindberg Corp., Chicago; Charles Yaker, president and chief executive officer, Howmet Corp., Greenwich, CT.

Contents

So You Want To Be a Consultant

The game plan is simple.
Lend a helping hand to engineers wrestling with the idea of going into consulting.

- Advise them authoritatively and from many points of view on what they ought to know before making up their minds.
- Give those who decide to try it a survival kit.

Help is almost exclusively in the form of a panel of 45 engineering consultants with collective experience of 516.50 years. Average experience is 12 years. Least experience (reported by three) is one year. Two are almost tied for "most." One, 45 years; the other, 50+ years. Forty of the 45 panelists consult on their own, one has a partner, and four work full- or part-time for consulting firms.

A second, smaller panel made up of clients in industry speak out on topics like:

"How do you locate consultants when you need them?"

"How much importance do you attach to resumés in evaluating candidates for consulting jobs?"

"Where do you rate ability to communicate (orally and writing)?"

"Is ability to get along with your people a key asset of a consultant?"

"Does size of fee charged by a consultant ever become a hang-up for you?"

As a by-product of the combined input of the two panels, a three-dimensional portrait of the engineering consultant emerges—another yardstick for the engineer who wonders whether he is suited for consulting and whether he would like it.

Does he, for example, totally enjoy being an engineer? Is he a dedicated professional who makes a point of keeping up with the state of the art? Is he problem oriented? Results oriented? People oriented? Would he take a job without pay if the challenge was great enough?

Members of both panels are named in the preface. Consultants are further identified in a directory at the end of this book (Chapter

2

30). Information in Chapters 1 through 24 was supplied by consultants, while that in Chapter 25 came from clients. Sources of information for other chapters are cited according to standard practice.

In the text that follows there is no attribution to individual panelists, but it should be understood that unless otherwise indicated, 45 experienced engineering consultants, not the author, are the sources of the experiences, opinions, and prejudices reported. The same applies to statements such as "consultants say" or "43 per cent report" or "it is common practice." These reporting techniques were also used in playing back the experiences and opinions of clients.

There is a small overlap between Part I of the book, which covers what engineers ought to know before making up their minds to leap into consulting, and Part II, which focuses on know-how needed to set up and launch a practice. The reason: the former, in part, is by necessity a summary.

Basically, it's a world-class step from engineering to consulting. To succeed, in addition to being at least a very good engineer, you will be obliged to take on responsibilities which are foreign to you and which you may not be particularly suited for or care for. Consulting, like engineering, is a profession, but it is also more. As a self-employed small businessman, for example, you will be obliged to sell, to market, to keep books, to make estimates, write proposals, present proposals orally, cope with client personnel afraid you are out to get them, travel probably more than you do now, work longer hours than you probably do now, etc.

However, the pay, if you consult full-time, will probably be anywhere from two to three times better than you were doing as an employed engineer; and other benefits—such as being one's own boss with the freedom to pick what to work on and when, plus the job satisfaction that goes with the territory—are at least 100 per cent better.

Three fields are covered: consulting in industry; working for lawyers and insurance companies, including being an expert witness; and teaching part-time. Consultant panelists make specific references to legal work and to teaching. Otherwise, reference is to work for industry. Also, you will find some reference to consulting full-time, part-time, or as a hobby. Generally speaking though, the information is universal in nature and applies across the board.

Topics in Part II are loaded heavily in non-technical areas outside the normal scope of an engineer's experience as an employee. In setting up and starting a practice, the engineer who makes a go of it must acquire basic skills of the businessman.

In this new environment, familiar subjects like chemistry, metallurgy, and trigonometry are teamed up with not-so-familiar subjects like advertising and selling one's services; setting up an office; dealing with lawyers, accountants, tax consultants, and bankers; determining a proper fee schedule; deciding on practical terms of pay for clients; knowing when to ask for up-front money and how much; learning the do's and don'ts in writing contracts; wining and dining clients; bidding, estimating, writing proposals, giving oral presentations; sensing the advisability of joint deals with other consultants; and checking out credit ratings of possible clients.

Information from the panel on being an expert witness and teaching part-time is supplemented in separate chapters (26 and 27). Sources are employers of expert witnesses (lawyers) and part-time teachers (schools and other organizations offering such training).

Lastly, some loose ends are dealt with in Chapter 28—Social Security regulations, record keeping, taxes, and the legal liability of consultants.

Part I

What You Ought To Know Before Making Up Your Mind To Be an Engineering Consultant

Chapter 1

What Consultants Do

Overview

When an engineer takes the leap from working for a boss to being his own boss, he assumes a whole new, and probably unfamiliar, set of responsibilities and risks.

For example, now that you no longer can count on a regular paycheck, how far will your bankroll stretch before Job No. 1 appears?

For example, among other things, you are now your firm's star salesman and marketing man and must worry about locating clients. The bulk of them will be in three fields: industry, legal work, and part-time teaching.

Where Consultants and Clients Come From

"I was laid off but had four months to find another job. During the grace period I lined up six consulting jobs in industry—with companies I had moonlighted for when I was a full-time employee. This is how I got started as an engineering consultant to industry."

"My entry into consulting came a week after retiring from a position with a well-known federal agency in Washington. A large firm hired me to help prepare a proposal for a government contract. They didn't get the job, by the way."

"I had no prior experience as a teacher [of engineering subjects] when a friend asked me to substitute teach for him while he was out of town on business. His trip, by the way, was for a job interview. He got the job, and I subsequently took over his class and have been teaching part-time ever since [1965]."

"A company being sued in a product liability case invited me to serve as an expert witness. I got the job through the recommendation of the company's technical people who were acquainted with my writings on the subject involved."

These examples give answers to three questions of practical concern to the engineer who hasn't made up his mind whether he is suited for or really wants to be a consultant at some future date:

1. How long should one be prepared to survive until the first client (and money) comes along?

8

2. Where do clients come from?
3. What is the scope of job opportunities? The comparative financial rewards in each?

First Client and Where He Comes From

How soon the first job comes along can depend upon factors like the state of the economy, how well one is known by reputation, and the extent of prior contacts in appropriate places.

Experience shows the waiting period may drag out from six to nine months, or there may not be any waiting at all—as when the neophyte has a client in the wings before his last day at work, or when he continues on the job for his former employer as a consultant, for example.

The first job may not be even remotely related to what one was doing; and it may turn out to be a fleeting or a long-term relationship with the initial client, as revealed by the following examples:

"I waited less than a month for my first client. The result was great. I was awarded a yearly retainer."

"I waited about one day. The client was a customer [paper mill] I did some free consulting for when I was an employee. You could say I converted a customer connection into a client."

"There was no delay. I continued the work I did before retiring for the same employer."

"I did not plan to be a consultant. I was asked to help on a steel mill problem about six months after I had retired."

"The day after retirement I was engaged as a research consultant for a steel producer, a half-time job that lasted eight years."

"After retiring, I did some soliciting for business as a consultant. The week after my first call I picked up a retainer on general problems from an electric products company which lasted for three years."

"When I started I had two consulting contracts that lasted about two years."

"I waited six months—for a job with the United Nations in Argentina. It was an excellent experience."

"It took me about one month. The assignment was making third-party weld quality judgments . . . worked out fine until they wanted me to do a lot of routine inspection."

"I had several offers to consult before I retired."

"I did not wait. Work came to me unsolicited."

"My first job was for the company I retired from. The next two were retainers from other companies."

"After one week I got an assignment for planning an extrusion facility for Zircaloy tube hollows. I also wrote the specifications for the equipment and directed its procurement. The facility has been in operation since 1970."

"I put off working as a consultant for three months after retiring; I could have started almost immediately."

"After two weeks of leisure I was asked to investigate the heat treatment of a railroad journal bearing. I eventually solved a problem MIT had given up on."

Another way of getting started—buying out a going practice from an engineering consultant—deserves only limited discussion. It has been done, but there is an almost unsolvable problem that stems from the nature of the profession.

You can buy a consultant's facilities and staff, but his reputation—his most prized possession—is nontransferable. There is a possible exception: a purchased practice has a chance of succeeding if the purchaser has worked with the seller long enough to have made a favorable impression on the clientele involved. An established consultant, perhaps a partner, is a more likely prospect than a beginner for such a deal.

Major and Minor Markets for Consultants

The mainstays are consulting for industry, serving as an expert witness, and teaching part-time. Actually, legal work goes beyond courtroom appearances. Lawyers, companies suing and companies being sued, along with insurance companies hire consultants to handle a variety of engineering detective work.

Consultants also:

- Consult for government at various levels.
- Consult for other consultants—usually as a subcontractor in a discipline in which the client consultant does not have expertise.
- Consult for testing and research laboratories and technical study groups.
- Consult for trade associations and technical societies in a variety of capacities.
- Consult for product designers.

Teaching goes beyond the standard classroom type. Consultants regularly conduct seminars and lecture on their specialties at technical meetings, conferences, and universities. Honorariums are typically offered.

Some consultants also own and operate testing labs.

Not all consultants work for pay. For example, members of the Union Carbide Retirees Service Corp. donate their time to small businesses and recover only their expenses. Other consultants volunteer their services to such organizations as SCORE (Services Corps of Retired Executives) and IESC (International Executive Service Corps).

Most consultants work with industry. Legal work ranks second and is growing in importance, while teaching is a relatively distant third.

Many consultants operate in two or three fields. Preferences of the panel for this book are as follows: industry only, 28 per cent; expert witness only, 3 per cent; industry/expert witness/teach, 30 per cent; industry/expert witness, 27 per cent; industry/teach, 12 per cent.

Consulting for government, particularly the federal government, is relegated to minor billing because it is the least favored among all markets. Excessive red tape and inundating paperwork along with low pay and slow pay are usually cited.

The engineering consultant in Washington is a small businessman dealing with big government. He is well advised to have the backup services of accountants and lawyers. Detailed records must be kept. Verbal agreements are to be avoided; written contracts are the rule. Contract language and terms should be reasonably simple and understandable. In work for the federal government, nothing should be written or done to arouse the curiosity of the watchdogs at GSA (General Services Administration).

Existing in such an environment drives up costs. A consultant comments, "We work on subcontracts for General Electric, Battelle, etc., but the red tape for direct government contracts is too expensive for us."

Quick Review of Going and Fixed Rates of Pay

The consultant typically determines his own fee schedule, with due respect for general practice in his area of expertise. Government, on the other hand, sets limits on what it will allow. The ceiling in Washington, for example, is around $40 per hour, or about half the standard rate in industry and legal work.

Consultants wise in the ways of government suggest a countermeasure in dealing with this set of clients. When jobs are put on a bid basis and the hourly rate is fixed and not negotiable, account for the time needed to cope with red tape and paperwork in estimating time needed to complete the project.

About slow pay, another consultant recommends, "Your checklist should include finding out who to call if payment is late."

Of course, some consultants are willing to put up with the trials and tribulations of working for government; but even these hardy souls exhibit the proud independence of a consultant in voicing their reservations. "I try to avoid government work. Otherwise, I simply follow the rules."

The majority position reads something like this:

"Don't plan on making money; do it for your country rather than for your bank account."

"I do only free work for them on occasion."

"I am a retired U.S. Navy officer and have had good and bad times on both sides of the fence. Generally, I suggest, avoid where possible. If it is patriotic, proceed. Otherwise, forget it."

Pay for part-time teaching is also on the low side; and as in the case of government, the client determines the fee. However, any differences seem to be offset by the intangible rewards of teaching.

As a group, engineering consultants are charitable and regard it their duty to their profession and to their communities to teach part-time at technical schools and colleges for the equivalent of $20 to $30 per hour, perhaps less. They can and do command at least $50 per hour or $400 to $600 per day for industrial work and at least $75 per hour or $750 per day in legal work.

Other schemes range from half-day minimum charges established by consultants (in one instance, $500 per half day for legal work) to retainers established by clients (usually on a per diem or monthly basis). Some consultants favor sliding fees based on the difficulty of the task. Some advocate a single fee for all work.

Chapter 2

The Quid Pro Quo

Overview

Anyone under the impression that consulting is a romantic easy street in comparison with the daily 8-to-5 grind as an employee has been poorly advised.

The pay can be better, but the hours are as long or longer; and the consultant inherits a laundry list of employerlike responsibilities. There is a chance that the would-be consultant may not particularly care for or be qualified to handle all of them.

Can he sell? Can he market? Can he make written presentations? How good is he at oral presentations? How is he at dealing with top executives? With client personnel who feel he is out to eliminate their jobs? Does he know the ins and outs of bidding? Of estimating? Can he do his own bookkeeping? Who will do the clerical work? The typing? Does he have a banker? A lawyer? A tax consultant?

Allied topics investigated include: the consultant's total workload; means of entry into industry, legal work, teaching; skills of an engineer that put him in good stead as a consultant in industry, in legal work, in teaching; consultants' feelings about each field; and finally, representative answers to: if one likes engineering, will one like consulting?

How Much Do Engineering Consultants Really Work?

Engineers, like other professionals, subscribe to the hard work ethic. The norm is working x hours per day at the plant or office plus x hours per evenings and weekends, keeping up with the leading edge of technology via reading, and attending conferences and seminars—in spare moments at such events, at coffee breaks, in hallways, at bars and restaurants, small talk centers on shop talk. This strong urge to share knowledge extends to giving technical talks and writing papers and articles.

A change takes place when the employed engineer converts to a consulting engineer. His passion for keeping up with the state of the art continues. Now, in addition, he exhibits great pride in a new-found capacity for self-determination—a freedom to choose who he will work for, what work he will do, and how much he will work as a consultant per day, per week, per month, per year. But you can't conclude if the consultant opts for four hours per day three days per week that he has deserted his traditional work ethic or that he has lucked into a sweetheart deal.

Total hours worked, even for the retired engineer who became a consultant, are probably close to what they were before—counting study, attendance at legal meetings, running the store, etc.

Actual time consulting is typically determined by how much income is needed for the consultant to support himself and his family in the manner they were accustomed to, or prefer to be. This is particularly true when the engineer is 100 per cent dependent on consulting for a living. His income is at least equal to what it was on salary, or double, or triple that total.

There is a built-in difficulty in trying to organize time worked by consultants into neat little stacks and give them labels like "high," "medium," and "low."

For example, a consultant may work a respectable 45 weeks per year, but on examination one learns a workweek in this instance averages 3.5 days, which adds up to 157.5 days per annum. What appears to be "high" actually belongs in the "medium" group.

Also, one must account for the fact that "amount of work" over a long period can be deceptive. A consultant's work tends to be sporadic—a day here, a week there, with time off in between. In addition, some part of the work comes to the consultant in the sense that he does not solicit it—another unpredictable factor.

Two other variables come to mind:

1. The engineer enjoys the option of spicing up his work with variety. In totaling time on the job, one may be obliged to account for a variety of occupations.
2. In addition to fulfilling continuing education requirements, the consultant is a small businessman with resulting administrative, etc. chores.

For example, an engineer who "works" 282.5 days per year (78 per cent of his available time) spends 800 hours/year consulting for industry, 100 to 200 hours/year serving as an expert witness, and 5 hours/week teaching. Those items account for 147.5 days/year. In addition, this man devotes 500 hours/year studying and 400 hours/year on administrative, etc. chores. Items that require an additional 125 days out of a year.

For what it is worth, it appears that a consultant who works 75 per cent or more of the time is in the "high" stack. The "medium" category starts around 40 per cent, while the "low" side is around 20 per cent or less.

Example of "medium": a consultant works 45 per cent of his available time—3.5 hours/week for 45 weeks/year—in industry, no time as an expert witness, and two to three days per year teaching.

Example of "low": one day per week with industry, no time as an expert witness, and ten weeks per year teaching.

Getting Started in Industry, Legal Work, Teaching

Consultants commonly graduate from industry. This is where the bulk of their contacts are, both in and out of their companies; this then, is where they normally get started as consultants. As a group, they probably are less prepared to gain entry into legal work and teaching.

Writing and speaking are typical ways of becoming known to talent scouts in all fields of interest to the consultant. Similar results are obtained via word-of-mouth recommendations from peers.

Other ways it happens in teaching:

"I volunteered to teach a course for the American Society for Metals . . . after lecturing within my company, offers from sponsors of seminars and universities followed . . . applied by letter after seeing an ad in a technical publication . . . I advertised in a technical journal . . . was volunteered by my boss . . . through recommendation of my plant manager to the local school board . . . by referral from a friend attending graduate school . . . through a neighbor who is on the staff of a junior college . . . I was the only person in the area qualified to teach the subject . . ."

A sampling of ways consultants get one foot into the courtroom:

"A lawyer friend with a product liability case in my field asked me . . . as a result of what I found in investigating a failure for a client, who subsequently sued his subcontractors and asked me to be his expert witness . . . through letters and resumés I sent to lawyers . . . through a discovery I made in research which was related to a suit involving an exploding bottle that put out an eye . . . I advertised and attended legal conventions where I met lawyers . . . I advertised in the yellow pages . . . as the result of a referral from a technical society . . . a company being sued sought me out . . ."

Engineer as Teacher, Expert Witness

The engineer is by nature a communicator, an explainer—attributes that give him a flair for teaching in the classroom or in the courtroom. In fact, consulting is often regarded as a form of teaching.

An engineer comments, "I feel I have taught all my career as a manager, executive, and presenter of technical lectures, etc." A teacher-consultant adds, "Teaching is difficult, but no more so than consulting, which is also teaching. Teaching prepared me, more than anything else, for consulting. I enjoy explaining things." A consultant specializing in legal work comments, "I get the feeling that doing a good job as a teacher is very much like doing a good job explaining a technical point to a jury."

Almost without exception, engineers enjoy teaching in the classroom, but there is an even split of opinion on whether it is "easier than" or "harder than" consulting in industry.

Teaching is demanding—more so, generally, than consulting in industry. Success requires a combination of knowledge, detailed preparation, much attention to the literature of the subject, and careful thought. Preparing for the next session can be a hassle if there is competition with concurrent consulting assignments in industry or legal work. Teaching schedules are inflexible and also can cause conflicts with other work, even prevent the teacher/consultant from accepting an attractive new assignment; and the chore of maintaining a fresh supply of examples that augment the text is among the problems cited.

What it takes for success in teaching tends to be something of a matter of personal preference. For example:

"A willingness to be fresh in your approach . . . a real interest in the students . . . a wealth of personal experience related to the textbook . . . enthusiasm . . . a teacher must become involved in the success of each student in his class if he is going to enjoy teaching and do a good job . . . teaching on the level of the student is essential, as it is to talk to the audience in presentations in court before a judge, lawyers, and jury . . . I believe the best teachers like to communicate and work at it . . . must enjoy explaining technical subjects to nontechnical people and be a ham . . ."

Pro's and Con's of Being an Expert Witness

A few engineering consultants enjoy the give and take of the courtroom. A few take the position that "it is just another assignment. I don't think in terms of like or dislike." A few "refuse to do this type of work." The remainder have some serious reservations about life in the courtroom, but they enjoy those aspects of the work similar to what they do in industry and in teaching.

As mentioned before, the term expert witness is shorthand for a range of legal work done for lawyers and insurance companies. Consultants investigate the technical basis of claims for both insurance companies and lawyers. They advise insurance companies on the merit of claims. They discuss the findings of their detective work with lawyers and give recommendations, join in the planning of courtroom strategy, prepare presentations for courtroom use, and go to law offices to give testimony under oath and adversary questioning for depositions for use as evidence in court. As required, they enter the courtroom to take the witness stand and give testimony under oath as an expert witness for either plaintiff or defendant.

Relative time spent in each area is indicated by an example from a consultant who devotes 750 hours per year to legal work: 600 hours to investigation, 75 hours to discussions and meetings with clients, and 75 hours giving depositions and testimony in court.

What some consultants like about legal work: "It is more difficult but also more challenging and enjoyable than normal consulting. I like the give and take of questions and answers, and explaining complex technical problems in everyday language. I like all of it . . . I like doing a good, honest, and thorough job, and being proved right by failure analysis . . . the challenge of reconstructing a failure in layman's language . . . planning presentations and strategy for our side . . . listening to the opponent's expert witnesses and planning interrogatories and strategy for cross-examination . . . I enjoy the game of wits with opposing attorneys and expert witnesses, but I detest sitting on hard courtroom benches waiting to testify . . ."

To the majority, the courtroom can be alien, unfriendly turf where their most prized professional resource—their reputations—is subject to attack and where pressure and tensions are created by lawyers who do not always operate in the logical manner of the engineer or show the same high regard for scientific fact and ethical behavior.

Critics of legal work charge: "It is more difficult because of the pressure of hostile attorneys and expert witnesses on the other side . . . very, very depressing to lose in a technically correct case . . . do not enjoy the apparent necessity of giving testimony about indefinite matters . . . do not like the tension in court and the gap that exists between scientific fact and legal practice; the aim of the scientist and engineer is to determine what did what; the aim of the lawyer is to get the most the law will allow for his client, guilty or not . . . don't like being badgered by attorneys . . . must be a true expert to survive . . . don't like the way lawyers try to get you to oversimplify or to bend your testimony . . . dislike the fear of inadvertantly saying something questionable . . . many expert witnesses are incompetent; some are knaves; some are both . . . legal work is not as objective as engineering; you have to combat illogical arguments and lies."

Good Money, Good Hours, No Office Politics

If you like being an engineer, chances are excellent you'll love consulting. There can be more job satisfaction, more self-fulfillment, more opportunities for personal and professional accomplishment. Consulting has special compensations for the retiree.

On the personal side, consultants enjoy the challenge of problems they have chosen to tackle, the stimulation of variety in their work, the sense of triumph that comes with solving someone else's problem, the feeling of independence and freedom that comes with being one's own boss, the rewards of good-to-excellent money, and ideal working conditions.

One consultant sums it up this way, "Yes, I would do it over. With my background, I am doing just what I wanted to do—having fun, helping people, making money. What else is there?"

As a professional, the engineer consultant has a strong sense of responsibility to help. One remarks, "It has been interesting, instructive, fun, remunerative, and I hope I have made useful contributions to the general welfare and to the profession."

Retirees find additional satisfaction in consulting. Says one: "As a retiree of 18 years working at my own tempo, I think it is prolonging my active life, and I hope to continue until at least 80 . . . it helps to make keeping busy easier, and life more useful and happier."

Chapter 3

Are You Qualified?

Overview

Guidelines that will help the engineer determine in what ways he is qualified to be a consultant are based on such considerations as:

1. Do you want to do it for the right reasons? What are the right reasons?
2. How attractive will your resumé be to prospective clients—work history, professional accomplishment, reputation of your employer(s), etc.?
3. Are you a registered PE?
4. Will your age be a plus or a minus?
5. Are you a competent communicator?

Key Question: Why Do You Want To Be a Consultant?

If you don't look beyond the obvious, "I was laid off. . . .I took early retirement. . . I quit my job. . . .I got fired. . . .I had to supplement my retirement income," there isn't too much to talk about. However, it's fair to say engineers become consultants for almost as many reasons as there are engineers. In the pursuit of this inquiry, one learns the reasons behind the reason engineers choose to become consultants. For example:

"I became a consultant to supplement my retirement income . . .and I enjoy the professional associations and problems. . . ."

"I was laid off from my job . . .had some 18 years of experience and had moonlighted a bit. . . .it was a natural transition. . . ."

"I was forced to retire by an unexpected ruling of the board of directors . . . .I had no desire to stop working, which I enjoy. . . ."

"I wanted to stay active after retirement at 65. . . .I was influenced by the fact that earlier my nephew had become a consultant to support his family and made a go of it . . . ."

"Before retirement, I did a lot of free consulting for many, many years . . .I needed the impetus of retirement to launch into a new career . . . ."

"Did it for fun after retirement . . .didn't need the money . . ."

"Always planned to because my first job was as #2 man for an established consultant . . ."

"While a faculty member at a university I consulted on the side . . . going into consulting after retirement was a natural progression . . ."

"I quit my job to become a consultant . . . I was bored . . . no new challenges . . ."

" . . . hate working from 9 to 5."

" . . . because of interest in my field of work."

" . . . wanted freedom."

" . . . to change my life style and to continue contacts with associates."

" . . . I enjoy teaching and being an expert witness. Also enjoy making money."

" . . . took early retirement at 61. Felt I wanted to continue work in my field. Also wanted to remain active on technical committees and in professional societies, and wanted to earn money to defray travel-associated expenses."

" . . . wanted to keep active and productive, and make a few bucks . . . no other work was available . . ."

" . . . becoming a consultant was suggested by several of my first clients while I was still an employee."

" . . . to continue a 50-year association with my profession and the hope of contributing to it."

" . . . I was beyond the usual age of employment (70) when I started to work for lawyers."

" . . . I wanted to be a consultant for 30 years . . . the fun of new jobs makes it all worthwhile . . ."

" . . . I became a consultant to avoid moving. The office where I worked closed. I felt I could make out OK as a consultant and realize a higher income. Also wanted to be free of the restraints of a large corporation . . ."

Three continuing themes run through all those comments: "I enjoy my work; it's fun. I enjoy the independence of picking and choosing clients, what I work on, and how much I work. I also have a strong need to be useful, to be productive, to contribute to my profession and to the public good."

Reputation: That of the Consultant, That of His Former Employer(s)

Reputation is what the consultant has to sell. "All of my jobs," says one, "come by word of mouth from my past performance in the metalworking industry." Another adds, "I live solely on my reputation. I do not solicit business. Nothing advertises as efficiently as satisfied clients." Both men are established consultants.

The beginner, on the other hand, has no record of accomplishment as a consultant and must rely largely on resumé-type information, such as job history and previous employers. Opinion varies on the relative merits of the two items as circumstantial evidence of reputation.

You hear:

"Job history is very important. Reputation of previous employers usually isn't a consideration"; or "I worked for a large oil company. My diversified experience and the reputation of the company have been very valuable"; or "My job history with International Harvester, Union Carbide, and Rocketdyne has been very important to me"; or, "It creates a favorable impression if you have performed for employers with reputations as front-runners with integrity."

Previous employers have particular importance in legal work. A consultant advises, "One's job history plays a key part in court as an expert witness." Another says, "Lawyers want a good line of experience. No one knows who my consulting clients are."

An additional source of reputation for both the beginning and established consultant should be noted: the reputation you have established in your industry among your associates and peers.

Once the beginner is fully launched in his new orbit, his record of performance takes over as the chief source of his reputation.

Of course, resumé information is also used effectively by the established consultant in the following manner: "I consider it very important as an opener with a prospective client, but otherwise, performance of the consultant on the job is the sole factor."

The Need To Be a Registered Professional Engineer

The engineer who has not had to be a PE as an employee should check state requirements if he becomes a consultant. Holding oneself out as an engineer in a business sense may require registration. In addition, state requirements for registration of consultants should be checked out.

The question of how much being a PE actually benefits the consultant in his business gets a mixed response from the panel for this book—60 per cent are registered PE's, based in Alabama, California, Florida, Illinois, Indiana, Louisiana, Massachusetts, Maryland, Michigan, New York, Ohio, Oregon, Pennsylvania, Texas, and Ontario.

On the one hand you hear:

"I have not had a client who required a PE . . . being one helps with credentials, I find, but not in the field . . . being a registered PE is a very slight advantage to me . . . my state does not require registration; the biggest advantage I see is in being able to answer, 'Yes,' when I am asked by a lawyer during a trial . . ."

On the other hand:

"You never know who will be impressed or demand it . . . it is a definite advantage with some prospective clients . . . it adds prestige . . . it probably helps the most when one does not have a Ph.D. . . . in almost every court appearance, being a PE is used as official documentation of your qualifications as an expert witness."

Age Takes on New Meaning in the Consulting Profession

A unique thing about consulting is that it extends the accepted age of employment beyond the age of retirement and puts a premium on age comparable to that assigned to youth in industry. In fact, in consulting, being a knowledge/experience-based profession, youth is viewed as a disadvantage.

The thinking is stated, "Age implies experience and judgment, unless, of course, the person is senile."

"In legal work," says another consultant, "the youth of a consultant may be a detriment, but age is never. Clients and attorneys seem to believe (sometimes erroneously) that wisdom, experience, and capability come with age."

Numbers are put on "too young" and "too old." In one view, "Too young means under 30 and lacking experience in engineering and in the business world. Too old is, say, over 70 when one is no longer current with the knowledge of his discipline, can't put up with the physical demands of travel, for example, and is bothered by the normal pressures of working."

Health is an implicit consideration. For example:

" . . . age enters the picture when travel is extensive and hurried . . . and hazardous physical exertion is required . . ."

"I will have a physical problem when I am no longer able to climb a pressure vessel ladder."

Age, of course, is relative. A consultant who is 82 would like to continue until 90. One who is 74 comments, "If jobs are available, I get them." Pragmatically, you are not too old if you are still alert mentally, up-to-date technically, and in shape physically.

A Consultant Is Expected To Be a Communicator

"Writing should be precise and to the point. Speaking effectively means being understood."

This quote flags the importance practicing consultants attach to the ability to write and speak effectively.

"Both writing and speaking are used to transmit your ideas. In many cases a written document is what my clients buy. In other cases, they will want a verbal presentation."

"Writing skill is essential to quality consulting. Same for speaking."

"Ideas and the expression of ideas are the stock in trade for the consultant."

"It's important to communicate in a language nontechnical people can understand, whether written or spoken. It is very easy to be too technical and lose them completely."

"Both writing and speaking skills are absolutely necessary, but beware of lecturing clients."

"I have become highly dependent upon a word processor. It helps me write good reports."

"Ability to speak extemporaneously is especially important."

"Both writing and speaking are quite important . . . to turn out first-quality reports and to be persuasive on the witness stand . . ."

Some consultants give writing and speaking even higher priority.

"If a person can't do both, he is severely handicapped as a consultant."

"Writing effectively is the key to your most important product: your ability to communicate your knowledge to others. The same applies to speaking, especially in legal testimony."

"My advice is: if you can't speak or write in the good to excellent bracket, stay out of consulting."

"Effective writing is one of the most important facets of consulting. Speaking effectively is only slightly less important."

"Forget about being a consultant without such basic abilities."

"I don't see how you can be a good consultant if you can't sell yourself and write a good report for the client."

Chapter 4

Profile of a Consultant

Overview

In writing specifications for an engineering consultant, one is bound to take into account his professionalism, the value he places on reputation, the pride he takes in his freedom from restrictions placed upon payroll employees and the accompanying gain in self-determination.

One could conclude that the bottom line will read something like, "rugged individualist," or "maverick," or "nonconformist." However, one must remember the other side of being a consultant—his role as small businessman. In this mode he observes contemporary thinking and practices subjects ranging from public relations and marketing to diplomacy in dealing with the client's top brass and employees down on the production line.

How Consultants Describe Consultants

The profile of an engineering consultant includes the following items:

- He is sometimes quiet, sometimes outgoing.
- He is a good talker, but a better listener.
- He is problem oriented and people oriented.
- He is thick skinned, takes command on the proper occasion, is well organized, modest, concentrates easily, is quietly confident, well dressed, punctual, tends to be a conformist, has good work habits, is an innovator, is diligent, diplomatic, a good salesman, has a sense of urgency and feels a need for accomplishment, likes the people he works with, and works well with individuals and with groups.

At this point it is worthwhile to supplement the foregoing macroview with microviews on two topics: views on conforming to current customs and styles and the somewhat related specialist/generalist issue.

The Consultant Tends To Be a Conservative

The consultant has a conservative dress code which is highly symbolic. "Neat and well groomed" are taken to have such peripheral meanings as "sincere, businesslike, professional, industrious, successful, competent, honest, and knowledgeable."

The extremes of nonconformity are to be avoided:

"I wear neat, conservative clothes that are in style—not a ten-year-old suit, beaten-up shoes, shirt collars and ties that are out of style."

"An engineer should have a professional appearance. Tie and coat or conservative suit."

There is no need to look affluent. Proper dress is defined:

"One should be well groomed without overdoing it, being phony."

"I dress for the occasion, even don overalls if I am working in a dirty area."

"A tidy appearance should suffice. To do the job you may have to put on overalls. But they should be clean at the start."

"During the first 40 years of consulting, I always wore a coat and tie. Last year I adopted summer dress and the informality has continued."

The judgment implied by those comments is reflected by the following: "I drive a $15,000 car, and wouldn't drive a $50,000 car. Clients wouldn't go for that, I feel."

Thinking on "what clients go for" supplies the rationale for the consultant's dress code:

"You must reassure clients of your knowledge. You should not be too casual in dress or in manner."

"Personal appearance is important to me and probably to my clients—simply because it indicates my respect for them and the job."

"Neither a slob nor a dude is likely to sell himself widely enough to make a living."

"Appearance is probably most important at the outset, but one should look and act like a professional rather than one who is affluent. Some companies have dress codes for executives and managers. The consultant should learn them and conform if possible."

"One should dress conservatively, especially in court where a jury might be influenced by appearance."

"Personal appearance is quite important in your first meeting with a client."

"Most clients prefer to deal with conservatively dressed consultants."

"An affluent appearance might be a disadvantage in court."

The Consultant: Specialist or Generalist?

The specialist vs. generalist issue is bound to come up in any discussion of our general subject, even though the end result may be of uncertain value. Perhaps a necessary peace of mind that cannot be otherwise obtained is gained from taking part in the exercise.

In any event, "Is it better to be a specialist or a generalist?" "Is it necessary to lean in both directions?" "Does it really make any discernible difference?" All seem to be fair questions, and knowing the consensus, or lack of it, has at least some philosophical value.

The outcome of the vote is preordained. No decision. A tie. All conceivable positions are taken. The value of the general discussion is in its documentation, the explanation or justification for a given vote.

First, the vote for specialist:

"Unless one is a specialist to some degree, I think he is useless. But jack of all trades? Probably master of none."

"My clients' problems have all been specific and required a specialist."

"I take jobs only in my special area, such as heat treating and mechanical engineering. If I am not qualified, I turn down the job, or call in qualified consultants to assist me."

"Depends on what the job requires."

"You must specialize for recognition in the area in which you serve."

"The generalist can overlook clues that properly define the problem, which in turn can produce an erroneous conclusion."

"In court cases I must qualify as an expert. This clearly means specialization. In most cases the consultant is expected to be a specialist."

The vote for generalist:

"Generalist, most definitely. Every job involves 30 to 50 per cent technical knowledge, plus a variety of knowledge, like costs, availability, processing of new materials. Every company has several problems."

"After 34 years as a superalloy specialist, my consulting is of a more general and fascinating nature. Off the record, I consider it the playground of metallurgy and enjoy 90 per cent of it."

"Generalist with specific areas of experience in failure analysis, welding conditions, and material processing."

"I consider myself a generalist, and my resumé outlines my areas of competence."

"Generalist, but know about specialties to ask the right questions."

The third position, "you have to be both," has as many advocates as the other two:

"It's essential to perceive what is needed—usually this means being both a specialist and generalist."

"I work both sides of the street."

"Both, I think. I have been a metallurgist and construction engineer for 33 years and have dedicated my life to it. My personal experience through pure luck has allowed me to be both generalist and specialist. A consultant needs a broad base to be successful. I guess I am a generalist after all. My specialty is stainless steel."

"I am a specialist in metallography but a generalist in that broad discipline."

"There is a need for both specialist and generalist. It is an advantage to be a relative generalist."

"I am a technical specialist, plus a management adviser, plus an internal relations improver."

"I am first a specialist, but one must be a generalist at times."

"I am a generalist in some areas and a specialist in others. I see the need for both. Consulting parallels the medical profession in this regard. If a client can clearly see his problem, then he should hire a specialist for the job. But generally someone with broad experience is needed to ferret out the problem. Then a generalist may be required."

"I regard myself as a generalist with skills as a specialist in selected areas."

"There is a place for specialists, but I personally prefer a generalist with broad experience."

"There are advantages to both. You are what you are. I have some specialties, but my strength is in the GP nature of my practice."

"A specialist must also be a generalist if he is to become a consultant. He must have a specialty, but be knowledgeable in related fields."

What's the outcome?

Suffice it to say, success does not seem to hinge on membership in any of the three camps. All of the panelists are successful. Perhaps this is what is to be learned from the discussion.

Chapter 5

More Risk Than Being an Engineer

Overview

"If you do not have intestinal fortitude and do not have faith in yourself, do not entertain the idea of consulting for a living," advises one consultant. In part he may be trying to discourage potential competition because consulting is a business, and competition is one of the normal risks of doing business. But the main point to be made is that the potential consultant must face up to the fact that there are night and day differences between being a payroll engineer and a consulting engineer. You must be prepared to accept an imposing list of new risks and new responsibilities as the quid pro quo for the benefits and perks of an entrepreneur.

What It Means To Be a Small Businessman/Engineer

"Beyond expertise," a consultant advises, "one needs to be a businessman, which is an altogether different ball game from drawing a check each pay period from a company. One must first of all be able to sell his services, then do a first-class job, then accept the very important step of collecting fees. Finally he must handle the accounting and tax jobs.

"Selling is difficult for me and perhaps for most engineers. But it has to be done even for the most expert and is a sort of career in itself.

"Consulting is also damned hard work. If a company is paying you $500 plus per day for help, not much time will be spent sitting around drinking coffee and telling stories. So a person must be in pretty good health and willing to work hard.

"Finally, as in business, there is competition. I am a full-time consultant; in addition, college professors consult part-time; and some retirees consult part-time to supplement their retirement incomes. This means consultants competing for jobs range from men having to make at least $300 per day plus travel expenses down to retirees willing to work for $100 per day or less to have something to do or to supplement retirement incomes."

Webster defines a sinecure as an office or position having emoluments with few or no duties. Consulting is no sinecure. In bolting the ranks of payroll employees, the engineer takes on an assortment of new risks and new responsibilities in addition to gaining benefits previously not available to him.

Almost to a man, for example, panelists emphasize that some risk is unavoidable in running one's own business. Two of the biggies have been cited: the generation and collection of income. Additional considerations stem from potential liabilities incident to doing business. Questions arise. Should the consultant take out general liability insurance? Is he ever required to post bond?

Opinions and comments on these and other considerations follow. But first a general statement of benefits and a couple of exceptions to the consensus.

Benefits, of course, can be substantial. The prospect of managing one's own work and time along with much better pay is extremely enticing. Other benefits among those worthy of note are in the legal and tax areas.

A consultant comments on the latter, after stating, "Yes, some risk is unavoidable, but risk can be kept to a minimum. Actually, you gain several benefits as a consultant, including tax benefits via an office at home, business use of car, and tax writeoffs on equipment like cameras, typewriters, and office furniture."

Another consultant explains his low-risk approach, "Mine is a no-risk business. I work only for truly reputable clients. My overhead is low." Amount of risk seems to be tied to amount of consulting activity. In the words of a consultant, "I don't feel I am taking any risks. Consulting is only a hobby for me."

An Assortment of Opinions and Comments

Being a consultant vs. being a payroll engineer:

"You have to work much harder and longer hours."

"I have learned more about business than I have about consulting. I told my wife at last I know what I am. I am a salesman."

"It is totally my show now; whether I make it or not is based on my own abilities, etc."

"There are two principal risks. One involves liability. The other is not getting paid or totally paid by clients."

"I would guess 25 per cent of my expenses are incurred by a client who asks me to investigate a problem, and the project is dropped after I supply the initial information."

"As a consultant I take all the risks, especially those connected with bill collection."

"As an expert witness there is some danger of not being paid if your client loses the case."

"You are definitely running a small business and should realize you run the risk of liability suits."

Some risk is unavoidable:

"There is some risk, of course; but with proper planning, the risk is no greater than that in operating a well-run company in the retail or service areas."

"I worry about errors and omissions."

"There is some risk, but it should be minimal. Insurance might be the biggest cost." Another consultant adds, "Nothing ventured, nothing gained."

Views on insurance and bonding:

There are marked differences of opinion on these subjects. One group takes the following positions.

"I have never seen the need for insurance or for posting bond. In legal work I am involved in product liability cases, but not in the sense they are my products."

"Have not purchased insurance or put up bond in 13 years. About 70 per cent of my work is with product liability and/or accident cases."

"I do not carry insurance. I have never put up bond. Have never been involved with liability."

"Insurance is optional. I have had the client take care of my bond."

"In some instances insurance and bonding may be necessary; but to date I have not done so . . . it depends upon how much one does."

On the other side you hear:

"You need general liability insurance, and I am going to buy product liability (error and omission) insurance."

"Insurance against being sued is desirable . . . "

"Insurance is necessary. I have never been required to put up bond."

The two comments that follow present overviews on the insurance/bonding issue.

"The need for liability insurance is a difficult item to tie down. Some companies include consultants in their policies. Others do not. (This may be a point of discussion at contract time.)"

"When the consultant makes materials recommendations for chemical applications, for example, serious liability considerations must be addressed. I hear the American Chemical Society has a liability insurance plan for consultants . . . "

Another consultant adds, "Liability insurance is almost as expensive as medical malpractice insurance, and is affordable only by full-time consultants with a very wide practice and large fees.

"I have been nervous about this problem. One of my clients manufactures oil tankers, and I worked on a number of different corrosion problems that put tankers into dry dock at a cost of $10,000 per day, and was informed if my solution did not solve the problems I could be held liable for dry dock costs in the future. Several years have passed, and no repercussions—to my relief."

Financial Aspects of Getting Started

Novice consultants have typical financial problems. They probably need a nest egg to fall back on; they may find it necessary to borrow money.

Getting started means much more than hanging out one's shingle and having enough money to get by on until the first client materializes. Financial requirements and problems at this stage cover a broad front.

For example, after you have clients, you will have lab charges that must be paid before you are paid by the client; you will have travel expenses that must be paid up front; you may want to purchase office equipment or lab equipment; you will probably want to supplement your reference library; you will probably want to consider converting your health insurance from a group plan to another plan or buy liability insurance; you will probably have office costs, such as an answering service, and secretarial costs to pay whether or not you have clients. Most importantly, states a consultant, "You will have to get used to a fluctuating income and be prepared to weather business down cycles and recessions."

Experienced consultants advise:

"Accounts receivable can strangle you, so insist on travel, per diem, and lab expenses up front. Never give a report without a partial payment."

"Keep careful books."

"To get prompt service from testing laboratories, pay their billings promptly. The same applies to consultants who are helping you."

"Do not let your plastic credit card get out of hand. Do not let your accounts payable for services or materials put you in a deep hole. Do not allow your own billings to go unpaid. Simply stay on top of your clients."

A potential benefit is cited: "Watch for tax deductibility of business development expenses in getting started."

Views on the necessity for a nest egg in getting started range from "No" to "Yes" and "It depends."

The No's are in the minority.

For example, "I did not have a nest egg. All of my bills are paid; I have $5,000 in my business bank account; and I have $10,000 in accounts receivable. I feel secure . . . that covers up to six months of living expenses."

For example, "I have no lab facilities. I subcontract all my test work, such as metallography and chemical analysis, to my former employer's materials lab. I get excellent services at moderate prices."

"It depends" comments go like this: "It depends upon circumstances; the dry spell can be long, and after that you may not make enough at first to get by on. I spent $11,000 in the first nine months and made $1,600."

Those who feel a nest egg is necessary tend to be fairly specific.

"For those without a reputation as an expert in some field, it may take as long as two to four years to get established. Cash flow is a problem for them."

"Unless you have gilt-edged clients at the start, you need enough money to last for two or three months."

"You should have two bank accounts when starting if consulting is to be your sole source of income. Don't try to start from ground zero."

"The first year will be tough."

"You need a nest egg if you do not have a sure source of immediate business. Even if you think you do, you may find the jobs aren't as immediate as they sounded earlier."

"One should be able to support himself for three to six months without a consulting job."

"It generally takes about three years to become established and switch from red to black ink."

"One should have sufficient reserve to smooth out irregularities in jobs . . . to avoid charging too much at the start and scaring off clients after one assignment."

"One must have credit or funds to permit travel to clients and to attend association meetings and professional conferences and seminars so one may keep up to date."

"One should not start consulting unless he has a nest egg to pay for expenses until he is paid for a job."

A couple of tips:

"Auto, office, travel, and secretarial costs are among those that must be paid before you are paid."

"Don't spend money before you get it. The brightest prospects often fall through. And don't do too much work without charging for it."

One consultant reports, "I have never had to borrow money. I was in the black the first month." He is the exception. However, as a group, consultants tend to be conservative in their borrowing practices.

You hear: "I borrowed to purchase expensive lab equipment and a word processor."

"In the beginning there are times you may need to borrow."

"Sometimes I find it necessary, but not for more than one or two months."

"Over $6^1/_2$ years I have borrowed money twice, in small amounts."

Borrowing from one's self seems to be a popular practice. "Cash flow is always a problem," says a consultant. "I have borrowed from savings several times, but I have not had to take out commercial loans."

"I borrow only from myself to promptly pay billings from testing laboratories and for other technical assistance."

"I often borrow from savings."

Chapter 6

Stress, Hours, Travel, Mistakes

Overview

You may be under less or more stress as a consultant than you were as a payroll engineer. On the "less" side you hear such explanations as, "I have the freedom to choose what I work on." On the "more" side you hear, for example, that there is more pressure to get a job done. Also, a full-time consultant will put in at least as many hours per week or per month as he did as an employee. But hours will tend to be more irregular and the frequency of jobs more sporadic. Travel (local and long distance) averages about nine days per month for the panel. Consultants, of course, are not omniscient and do not pretend to be. They also make mistakes and do not always come up with the sought-after result.

Slightly Less Stress Than Being an Employee

As an uneducated guess one would surmise that being a consultant is more stressful than being a payroll engineer. However, the vote of the panel was almost a toss-up, with 48 per cent reporting "less stress." The most common explanation boils down to, "Usually there is less stress because I can pick and choose only those jobs I feel comfortable with and enjoy doing. If a job appears to be overly stressful, I turn it down."

Or, "I am under no stress at all as a consultant, or at least I don't feel I am. This probably results from the fact I work as a consultant only part-time, and I am able to choose the jobs I work on. Meeting specific time schedules has never been stressful to me."

The consensus of the "more stress" group centers on "meeting specific time schedules." A close second goes to "fear of being wrong."

Some representative comments:

"There is more stress primarily because of the urgency of the job."

"Fast service is often expected of a consultant, or they would not be called on for their assistance and specific knowledge."

"In consulting you usually don't know the individuals well, so you realize it must be done right the first time, and this causes you to be very careful."

"You probably have to be more sure you are correct every time you make a suggestion."

Other reasons cited for "more stress" include:

"Yes, I am under more stress when I am in court."

"Working for someone else [other than an employer one knows well] is more stressful."

"Not so much stress, but a feeling of more direct responsibility."

"Yes, because you are representing yourself as a one-man company."

"Yes, there are more unknowns now, and I must address all facets of a business. It all depends on me now."

Some consultants under "more stress" recognize offsetting compensations:

"There is absolutely more stress, but the rewards are greater, both in satisfaction and in money."

"Yes, but I like it that way."

"Stressful conditions do not last as long as they did when I was an employee."

Consultant Vs. Payroll Engineer: More or Less Hours Worked?

Because the panel is made up of both full-time and part-time consultants (the latter are typically retirees), there is about an even split on the question: "Do you work more or less hours than you did when you were on a payroll?" The result is predictable: full-time consultants "more" and part-time consultants "less."

There was one surprise in the ballot: both sides agree "hours are more irregular than they were before."

Some comments:

"When I am busy I may work straight through to completion, putting in a 16- to 18-hour day."

"I work up to 60 hours a week. There are many details to take care of: bookkeeping, insurance, filming, organizing, and then there are interruptions."

"You have to take assignments as they come. I sometimes work seven-day weeks in preparing for a court case."

"The amount of work varies with business conditions. In a boom, I work five to six days a week, 10- to 14-hour days; in a downturn, I may work two or three days a week, 8- to 10-hour days."

Among panelists responding, the average workday is 7.7 hours; extremes range from 3 to 18 hours.

Average days worked per week is 4.3, with extremes of 1 to 2 to up to 7.

The average workweek is 30 hours; extremes are 8 to 60 hours.

Typical comments of retirees:

"I did not retire from full employment to go back to work full-time. I may work only a couple of days per week and put in a total of only 10 to 14 hours."

"I work less than I did before retirement. Much of my work as a consultant is in my office."

"I put in about a 50-hour week: 30 hours of consulting and 20 hours as a volunteer—church, social agencies, as a member of technical committees, and as an attendee or speaker at technical conferences."

Some consultants see no difference. One put it, "I work 60-hour weeks, but I always did—10- to 12-hour days, five to six days a week. That's because I enjoy what I do."

How Much Engineering Consultants Are Obliged To Travel

The need to travel, like workweek length, varies widely and is unpredictable. The panel's total travel time per month is about nine days. Travel by car, mostly local, averages about 12 days per month, while the average for travel by air and train (the latter to a limited extent) comes out to about 4.5 days per month.

On the high side, total travel per month runs up to 30 days (mostly by car and local); also, "24 nights per month away from home, with almost all travel by air"; and "21 days per month—20 by car, 1 by air."

On the low side, travel per month ranges from "seldom" and "rarely" to "one to two days—usually by air."

Most travel by car per month is 30 days; other high numbers include 20 and 15 days. With few exceptions, this is local travel.

Some comments:

"My travel is very variable. I am generally away from home three to four days a week, which I am trying to reduce. In the Los Angeles area I use my car; in other areas of California I travel by air if the airline schedules allow."

"My travel per month varies from one to two days to a week or more. Most of my clients are local or within commuting distance."

"My travel time varies from year to year. Most of my business has called for travel."

"Last year I traveled for three months—140,000 miles by air and 40,000 miles by car. In the first five and one-half months of this year I have not traveled by car. On the average, I log about 10,000 miles per year by air and 30,000 miles per car."

"I typically travel seven to ten days a month. However, my trips overseas usually run four to six weeks. I must take maximum advantage of my time overseas. Totally, I travel by car, rail, and air."

When Consultants Need Help, They Ask for It

Consultants are consultants because they know more than the average engineer. However, as a group, they are not cocky about their capabilities; they readily admit their knowledge and experience are finite and are quick to confess they frequently need help on the job and do not hesitate to go out and get it. "I haven't needed help so far, but I am not too proud to ask," says a consultant.

The need for help typically arises when a consultant learns he does not have all the required know-how to handle a job, or when he is not equipped to supply required support services, such as laboratory work.

Help comes from a variety of sources, including previous associates, protégés, industry, universities, fellow consultants, and friends. A consultant comments, "From 10 to 20 per cent of the time I obtain help from other consultants." Another adds, "I often ask colleagues in related fields to become associated with me, but by the job, not permanently." Another offers, "Good technical contacts are vital to doing a good techical job."

Help from commercial laboratories with skilled technicians is readily available. One consultant uses "support technicians in metallurgy and nondestructive testing." Another farms out "metallurgical photographic work."

Consultants Do Not Hit a Home Run Every Time

Making a mistake or not getting the desired result every time is not fatal to the consultant. Both case histories and pitfalls are cited by the panel.

"I was an expert witness in a case where an extension ladder failed when its hooks did not seat properly because ceiling height was restricted. A judge asked me to demonstrate in the courtroom where, I did not realize at the time, ceiling height was not restricted. The hooks seated properly three out of four times. My side lost the case. I should have realized the angle of support was critical and refused the judge's request."

"A push boat with a large tow of barges going up the Mississippi River lost all power; the current pushed the boat down the river, causing a collision with a plant on the bank, doing millions in damage. I found a leak in the freshwater tank into the fuel system and contended that the power loss was attributable to the manufacturer of the boat. I lost. The judge thought the captain could have avoided the collision by maneuvering the boat."

"A manufacturer asked me to help in heat treating some large, expensive gears. He wanted a rush job, and the gears cracked. I should have insisted on following standard procedures."

"I was hired by an attorney on a property damage problem involving galvanized sheet steel. In court, the opposing attorney asked questions about the manufacture of galvanized—an area of the subject where my knowledge is limited. My testimony, at best, was limited . . ."

In these examples, the consultant did not anticipate getting into deep water and could not avoid it. However, one of the cardinal rules is: *Do not take on something you are not qualified to handle.* "Sometimes," says a consultant, "an unqualified person tries to bluff his way through, pretending to be an expert when he is not. The results can be disastrous." Another way of putting it: "One should stay within his own field of expertise and turn down a job or call in other experts where required."

Chapter 7

Ethics, PR,
Irregular Work

Overview

The consultant had a very friendly, even breezy, telephone manner, but he listened intently, obviously thinking as his potential client talked. He took complete and careful notes in a clear hand on a legal-sized pad of yellow paper.

"Yes, I can handle it," he said, as he studied his appointment calendar. "Why don't we get together at 2 p.m. next Monday at your place."

After he put down the phone, the visitor in his office remarked, "I'm surprised, for an engineer you are quite a PR man."

The consultant smiled. "You have to be whether you like it or not."

The visitor went on. "You said you could handle the job solely on the basis of a few words over the telephone. What if you later find you aren't qualified for the job?"

The consultant did not hesitate. "The ethical thing in that instance would be to withdraw from the case."

Added the visitor, "It also occurred to me that you may have been able to give the man what he wanted on the spot, so to speak. Maybe your call isn't necessary."

"I do some of that," said the consultant. "It's called free engineering. You solve a problem without getting paid for it."

"Do you always get paid?" asked the visitor, "or do you ever have trouble with collections?"

"Sometimes," the consultant answered. "There's always a chance of encountering either slow pay or no pay." He added, "You also have to remember this tends to be a peak and valley business. You're either up to there in work or you are on hold."

Consultants Feel: "100% Integrity Is Essential"

Situation No. 1. "I turned down more jobs in the DC area than I accepted," recalls a retired consultant, "because many of the studies were completely worthless."

Situation No. 2. "I always refuse jobs when I feel I am not fully qualified. The only exception is where I feel I can do the job with help and get it."

Situation No. 3. "I occasionally work for two companies in the same industry. I inform the first of my intentions to work for the second and come to an agreement with them on what is proprietary information."

Situation No. 4. "My only attempt at being an expert witness was a fiasco. The client wanted me to use his answers. I refused."

The examples cover four typical situations in which questions of ethics arise. The accepted standards of conduct that apply have old-fashioned names like "integrity," "honesty," "objectivity," and "professionalism." The base rule is: "Ethics should be beyond reproach." In fact, some members of the panel declined to discuss the subject because "I will not divulge practices of my clients" or "I do not talk about shortcomings of peers."

Situation No. 1, giving good value, seems to be universally observed and is not cited as a problem area.

There is much more concern among consultants for colleagues "who will take any job even though they are not qualified." The practice is not epidemic, but consultants do report, "I occasionally get jobs that have been fouled up by other consultants who did not know what they were doing."

The problem seems to be more prevalent in court than in industry. The consultant just quoted adds, "In legal work I run into so-called experts who are not and who will make exaggerated statements and speak untruths."

Another side of the legal situation involves the laundering of technical findings or testimony. A consultant remarks, "A client may ask you to favor his point of view, and you must resolve this ethically."

The standard solution is suggested by another consultant. "Many clients want consultants to come to a predetermined conclusion . . . a desire to be resisted at all costs."

Another adds, "Occasionally in a product liability case my findings may be opposite what the lawyer's client wants to hear. One must be especially careful in such cases to be scrupulously honest, even to the detriment of the client." If this is intolerable, consultants often withdraw from the case.

The potential for conflicts of interest is natural because consultants tend to specialize in a given industry, but this does not preclude them from working for clients who are competitors. Clients typically have consultants sign secrecy agreements at contract time; and it is accepted practice of consultants to inform all clients involved of what is going on. A standard approach is to reach an agreement with the client on what is proprietary information. Further, consultants say, "I never discuss consulting work for a client with another client in the same industry."

As a group, consultants abide by strict rules of conduct and expect no less from their colleagues. Perhaps the level of the standard is characterized by the following comment.

"After a successful appearance as the expert witness for the plaintiff in a gas explosion case I proved was caused by leaks in gas company pipes, the gas company asked me to work for them in future cases. I refused, feeling it would not be ethical."

"If They Don't Like You, You Won't Accomplish Much"

"A loner probably doesn't make a good consultant," suggests an established consultant. He is among the majority who believes consultants should be outgoing, friendly, should like and be liked by the client and his personnel, and should have the respect and cooperation needed to ably get people to work well for him.

Although this aspect of the consultant's job has nothing to do with his technical competence, it is of at least equal importance, many consultants feel. One comments, "Technical expertise gets you the job in the first place, but it is vital to have good public relations."

The reason he gives: "It helps facilitate good communications with the client; i.e., it is easier to communicate from a positive position than from a negative one."

Among the tactics cited by consultants, three stand out:

1. Lubricating the various relationships one has with clients and client personnel one must work with.
2. Maintaining the charisma or image of a consultant/expert/ professional.
3. Selling one's self as well as convincing the client of the merit of an idea or proposal.

The following composite of comments states the majority view.

"You must be Mr. Clean, raise no questions, no doubts, be alert, sharp, and stay awake . . . in legal work you must convince both your clients and opponents and their attorneys that you are competent, cooperative, and that you handle yourself well . . . if they don't like you, you can't accomplish much, but that doesn't mean you shouldn't tell them when they are wrong . . . I believe in being friendly with everybody—supervisors and workmen—and being very open with them about what I think and what I can and can't do for them . . . very important for repeat business . . . sell, and sell hard, the staff engineers you work with . . . so much involves antipollution, environmental, human factors today one must be aware, alert, and public minded . . ."

The importance of personal PR is underlined by the following remark: "Having someone point out your reputation and achievements in advance is a big help in getting the respect and cooperation of the people you will be working with."

Another adds, "The image of the consultant as seen by clients and prospective clients is very important. The consultant must relate easily and well with those he seeks to help."

Another states the point in a negative way. "PR is necessary, but if you are well known, other people—including clients—will do it for you."

Some consultants place less importance on personal public relations. One even feels "it is not the consultant's job." However, he adds, "be careful not to be negative with clients."

Opinion in the former category runs, "PR is not of major importance; how much it helps varies. It is important if you are not dealing at a high enough level in a company."

Or, "My concern is getting the job done. Public relations is secondary."

There Is No Guarantee You'll Be Paid for Your Work

When an engineer takes the large leap from employee to consultant, he foregoes the security of a regular paycheck. Consultants usually expect to make money on each job, and most do; but as in any business, slow pay is common, and there always seems to be a chiseler or welsher.

Consultants cite other reasons for making no money or not as much as expected on a given job:

- Having out-of-pocket expenses that are not recovered.
- Giving free engineering in the hope it will drum up paying business.
- Doing a job for practically nothing because you like the challenge it presents.
- Helping a company in need with charity work.
- Being on the low side in estimates of time and expenses for jobs where contracts contain "not to exceed $" clauses.
- Donating some hours and some expenses on a job because of the desire to give good value.

Of course, under normal circumstances, a consultant won't take a job unless it looks like a winner. Experience varies, but batting averages in this respect seem to be decent.

"I have been very, very fortunate . . . have had only two losers in the past nine years."

"I make my full rate on 90% or more of my jobs."

"Yes, I make money 95% or more of the time."

"I have had only one loser, and no longer expect them."

Consultants usually charge on the basis of a daily fee plus expenses. But some out-of-pocket expenses may not be recoverable, as in the following instance:

"I have contacts in Japan for overrruns of silicon chips and PC boards; and when I am over there I may be asked by my sources to get quotes. Telephone and Telex can run $75 to $100 U.S. per inquiry. Say I make eight calls and find no takers. My sources will not reimburse me for these expenses."

Another example: "My out-of-pocket expenses and base hourly rate are always covered; but in some cases all overhead costs may not be covered."

Free engineering often occurs when the consultant is asked for a "little help," typically over the telephone. He gives it without charge with the expectation it may blossom into a full-sized job in some way or other. "At times," says a consultant, "I have not charged for small jobs, such as consulting by telephone, or for a friend." Another adds, "You must put in some seed-sowing time whether it is directly on a contract or not."

To a man, consultants are problem oriented. Problems are fun. Solving problems is even more fun. Some jobs are bid low intentionally just for the opportunity to have a crack at them. A consultant comments, "I usually intend to make money, and I do; but I do take some jobs just for the challenge."

Engineers are a charitable lot. "I expect payment on every job," says one, "unless I agree beforehand to donate my fee to a struggling company."

Another universal urge, "giving good value," is expressed a number of ways. "I tell some clients at the outset," says a consultant, "if I can't help, there is no charge. Then we take it by stages until it becomes go or no-go." Says another, "Several times I have given a client many more hours than I have charged for. I try to do the best possible job, even if it means exceeding the budgeted time or dollars."

Consultants are usually paid on the basis of time or per diem plus expenses. Much less common are clients who ask for bids on jobs or insist on an arrangement with "not to exceed" clauses that put a ceiling on what the consultant can charge and require him to estimate how much time a job will take and what his expenses will be.

"I expect to make money on every job," states a consultant, "but I have made less than I wanted to on six to eight 'not to exceed' contracts in five years."

He adds, "I experienced two no pays in one year, but still may get reimbursed. I have lost on a job only through nonpayment."

As in any business, the trouble with "slow pay" and "no pay" customers is that they can't be spotted until after the damage is done. One recourse of the consultant the second time around, should it occur, is to demand full or partial payment up front.

For Most, Consulting is Cyclical-Irregular

The cyclical or unpredictable nature of consulting work is another fact of life, like slow or no pay, to contend with. In fact, a consultant comments, "The volume of work is quite cyclical, as is payment for it."

Experience varies. To some, consulting is not intrinsically cyclical. "Not in my case," observes one. "People get into trouble any time of the year." Another adds, "In my case it is not. I have heard that it is, but in my case I have been fortunate in picking up business through former associations."

The majority is of another mind. "The rate of new case arrival fluctuates . . . often spasmodic . . . in the sense that most business is cyclical . . . cycles in the activity level are not predictable . . . ."

Perhaps "cyclical" is not exactly the right word. There are some seasonal influences, as in construction, or influences like "the volume of my work depends on military expenditures." But perhaps "irregular" is a better choice, or "cyclical-irregular."

"My business fluctuates," says a consultant. "I have not observed any logic." Another remarks, "I have worked 40- to 60-hour weeks for two or three weeks. I have been idle for four or five weeks."

Whatever the name for the condition, there are some countermeasures. Getting retainers from clients is one. Having a diversification in clients is another.

Chapter 8

Setting Fees, Calculating Profit

Overview

How do you establish a fee structure? Determine what to charge on a given job? Is there any way of knowing if you have a suitable fee structure?

Many practices, guidelines, rules of thumb, and opinions cover the first two items. The third is analogous to calculating profit margin and is iffy because consulting is a service, and cost of service is not comparable to cost of product.

You Sell Service, Not a Product

If consulting is more than a hobby and there is no other major source of income, it is important for the consultant to have at least an approximate idea whether he is overcharging, undercharging, or doing about as well as he can expect. In other words, is there enough left over after all expenses are paid? Conventional formulas for computing profit and loss don't work too well. There is no cost of product. Expenses related to a given job are usually recovered from the client. Overhead and sales/marketing costs for such items as an answering service, typing, a long-distance call here and there, and picking up the tab for an occasional lunch with a prospect tend to be on the modest side. However, there are some yardsticks, most of them empirical, that are applied by consultants.

Example: "I want to make enough to operate the business, pay living expenses, pay for my retirement plan and insurances, give some money away, buy equipment, and have $5,000 to $10,000 left over on an annual basis."

Example: "I want a substantial income and to be able to contribute significantly to Keogh. Incidentally, consulting firms strive for a 15 to 25 per cent profit margin."

Example: "The basic question boils down to: What is your time worth?"

Example: "I expect my earnings to be comparable to those of a top-level metallurgist in industry."

Example: Not all expenses are recovered at times, or fees may be reduced or not totally paid after a job is completed. In this context a consultant says, "If I net 75 per cent of the asking price, I am satisfied. On most jobs I net 100 per cent."

Example: "My personal expenses must not exceed 25 per cent of my income over a one-year period, so I can realize a 75 per cent margin before taxes."

Example: "$50 per hour (minimum of 8 hours) and all expenses paid will yield an acceptable profit on a job."

Example: "A 10 per cent profit after all of my expenses are paid, including personal expenses, and my salary."

Example: "My margin should be the same as that of an M.D. or an attorney."

Some consultants take the position, "I don't know how a consultant figures profit. I charge a fee of $500 per day plus expenses. What profit margin is that? My overhead is almost zero."

Another remarks, "There is no such thing as a profit margin for consultants. My net after all chargeoffs for income taxes amounts to 55 per cent of gross, with no markup on outside lab expenses."

39 Different Ways of Building a Fee Structure

The foundation for a fee structure is normally hours worked plus expenses related to the job. Beyond that, there seem to be as many options as there are on a new car. Thirty-nine different techniques follow. There is no particular order, because what is used seems to be pretty much a matter of personal preference.

1. "General knowledge of what lawyers will accept without protest."
2. "In the Washington, DC area, fees are geared to a Grade 18 [the top salary in Civil Service], which figures out to about $250 per day or about $31 per hour."
3. "I use whatever the market will bear."
4. "I compare notes with other consultants, and then judge what the market will accept."
5. "Variable, depending on the job, but basically a fixed fee plus expenses."
6. "I work for a publisher who has a fixed fee for reviewing books."
7. "I pay close attention to going rates and make sure I conform—if less is charged, there is an inference that you are lacking in confidence."
8. "By mutual agreement with the client if there is a retainer. Otherwise, my one-time fee is $500 per day plus travel expenses. My hourly rate varies from $50 to $75."

9. "My fee for a given client is established on my first job with him."
10. "I follow ACME [Association of Consulting Management Engineers] standards."
11. "My fee varies with the difficulty of the job and the personal inconvenience experienced."
12. "Primarily, the going rates for comparable service. Secondarily, by size and financial health of the company."
13. "About 2× or 3× the normal annual income of someone with my qualifications reduced to a five-day, 52-week basis. For example, $60,000 per year comes down to about $230 per day."
14. "Fee varies by the job, depending on its scope and level of expertise required."
15. "Generally, a basic fee modified by such factors as significance and difficulty of task; duration of task, and whether it is one-shot, continuing, short term, long term, etc.; travel requirements; home preparation requirements; nature and wealth of the client's business; one's reputation, expertise, etc."
16. "The going rate for such services in this area, i.e., about $500 per day for experienced consultants."
17. "One rate for industry, and a higher rate for product liability cases where litigation might become involved."
18. "My fee is based on the complexity of the project, length of job, and amount of money involved, such as whether it is part of a cost reduction project or a capital expansion program."
19. "My standard fee is about $450 per day plus travel and living expenses while I am on the job. If the job is long, the fee may be lowered."
20. "I use rates charged by consultants I worked with when I was employed as an engineer—there are going rates for different levels of expertise."
21. "Originally I based my fee on my former salary plus fringes. I have since made upward adjustments to keep up with inflation and with trends in fees."
22. "I check rates commanded by professional engineers and established lawyers."
23. "Going rates plus those of other professionals are the yardsticks, especially fees of attorneys and accountants."
24. "My fee is based on what my competition is charging."
25. "For commercial work, $300 per day; for aerospace and nuclear, $500 per day; for legal, $750 per day. In all instances, I recover all related expenses, too."
26. "I try to keep ahead of plumbers and nearly in a class with lawyers and medics."

27. "In setting up my fee schedule, I talked with a tax accountant and with other consulting engineers."
28. "On some jobs I quote an hourly rate. On others I quote a fixed rate. Generally depends on what appears to be the size of the job. In some cases, the client may suggest an hourly rate or a fixed price—particularly if the client is a government agency or a Department of Defense contractor."
29. "What the market will bear is my guide. I started at $500 per day; went up to $1,000; and now get $2,000 (for anything more than a half day)."
30. "My fee is negotiable and depends on the client. The standard is $500 per day, $70 per hour, and a minimum of four hours. For jobs longer than one or two days, I write a proposal and include estimates of such items as time required and expenses."
31. "Depends on the service I provide on a job. For example, I charge $600 per day plus 25 cents per mile for travel. For work in my office I charge $70 per hour."
32. "The rule of thumb in my field is $40 to $80 per hour plus expenses."
33. "I try to stay in the top 25 per cent of the going rate."
34. "Three times the daily amount I would expect if I were an employee."
35. "The rule of thumb for professionals like faculty members who consult part-time is 1 per cent of annual salary per day."
36. "At least 2× the daily salary for a similar engineer employed full-time by a major company."
37. "The minimum daily fee should be: annual earnings as an employee divided by 100."
38. "Charge what other metallurgical consultants do in my community."
39. "A beginner can get a rough idea of what to charge per hour by doing his first job for an acquaintance and when it is finished ask, "What am I worth per hour in your opinion?"

When the Job Is Over, What Should You Charge?

The bill you send a client after the job is over is normally based on the standard hours worked, plus expenses. However, as the following examples illustrate, other considerations may be taken into account.

"I usually charge $70 per hour, and a minimum of four hours or $500 per day, and try to keep the day at 8 hours. If I consistently work over 8 hours per day, I charge $50 per hour for anything over 8 hours. If I have a heavy work schedule, I consider charging travel time, particularly if the job took me away from other work I should be doing concurrently."

"I charge primarily on the basis of time and expense. Sometimes I may charge more if the findings are particularly significant."

"I charge for hours worked, but I sometimes also look at the value of the work and adjust accordingly. Sometimes I mark up outside charges by 15 per cent."

"I may not charge anything as a good turn for a struggling company. It is important to enjoy your work foremost and let the charge be secondary."

"On local jobs I charge for hours worked plus about 50 per cent of my time away from the office, including time spent traveling to and from the job."

"I charge by the hour for local work and by the day for out-of-town work plus travel expenses and travel time."

"I have established hourly rates and reimbursable expenses. What I charge on a given job is negotiated with the client up front."

"Where practical, I quote a fixed price."

"Charge the least on repeat business, and the most for jobs involving a visit to the client, a study or survey, and a report."

"I charge $300 per day."

"I charge by the hour."

"In general, you undercharge when you are getting started as a consultant. But remember, the client is not paying any fringe benefits—which can be 25 to 50 per cent of your terminal salary."

This tip for beginners is offered by a consultant:

"I quote my rate when contracted by the client to avoid any misunderstanding when the job is completed. If you have not worked for the company before, get a purchase order from them before starting."

Chapter 9

Financial and Other Rewards

Overview

You can make a great case for the financial rewards of being a consulting engineer, but such claims, standing alone, can be misleading even though they are categorically true. At least equal recognition must be given to "other" rewards. Intangibles such as being one's own boss and the job satisfaction one cannot get as a paid employee are examples. In addition, consulting puts one in a position to satisfy the consultant's strong sense of charity and his urge to make contributions to his profession and to society. The latter is documented by the following comments.

"I charge no fee, I look for no business. When I retired, I offered my services to southern California heat treaters on an expense-only basis. I felt that someone should represent the industry where and when needed to protect it from unrealistic and unusable heat treating specifications . . . for aerospace and military applications. The heat treating business has been good to me over the past 45 years, and I aim to do whatever I can to help it as much as I can."

More Money, More Freedom, More Job Satisfaction

A few consultants will tell you the financial rewards are really quite small, that "the best reward is the challenge of doing top-notch engineering work"; or, "financial rewards are not likely to sell the profession . . . the rewards are in being a social dropout with technical ability who values freewheeling above industrial security"; or, "for a retired person the profit is small by the time you pay back Social Security and your various taxes."

A few consultants take a middle position:

"My rewards as a retiree have been to stay involved, active, and have a feeling of accomplishment. Financial rewards are icing on the cake."

"There is freedom to accept jobs you enjoy. Financially, it is OK, but in the present economy [1983] I would not encourage anyone to enter it."

"Haven't had any financial rewards as yet," states an engineer who was obliged to take early retirement, "but consulting keeps me involved, always facing problems to be resolved, and always being with new people."

The solid majority, made up of full-time and part-time consultants and retirees, is downright enthusiastic about the profession, its rewards, and potentials.

"Lots of money, travel, exciting jobs, personal satisfaction in doing a good job, lots of diverse problems."

"A good paying profession; enjoyable variety in problems, people, and workplaces; satisfaction in helping someone out of a problem; pleasure of being thanked for something you were paid to do."

"Beyond the financial rewards," says a retiree, "there are also the ego rewards of solving problems, doing a good technical job, satisfying a client, knowing one is still productive and creative after retiring from a full-time position, and the knowledge that one can still be competitive."

There is less of a consensus on financial rewards than there is on those of the "other" type. On the money side, however, you have to account for the fact that our panel was a mixture of full-timers, part-timers, and retirees. Some, but not all, in the last two categories no longer have the drive and ambition they once had. To them consulting may mean that "extra money," or "supplemental income," or "I can make as much as I did as an engineer in industry" is a sufficient reward.

At the other extreme you hear, "The financial rewards are as high as one cares to work . . . and they can be very substantial . . . generally equal dollars for half the effort as an employee—up to two to four times what the average engineer makes in industry."

On balance, however, intangible rewards far outweigh financial rewards in number and are of at least equal importance to the recipient.

Several common themes run through the intangibles: freedom to choose clients and jobs and to establish one's work schedule; the challenge and job satisfaction in tackling and solving a diversity of interesting technical problems; the satisfactions in maintaining and updating one's technical knowledge; and the opportunity to make worthwhile professional contributions to industry and to the nation.

Some representative expressions of these values:

"Successful consulting should yield a very comfortable living, and a reputation as a highly qualified expert is an enjoyable experience."

"I waited to do it for 30 years. The fun of new jobs makes it all worthwhile."

"My income is considerably higher than it was, and I can write off expenses like business use of car, travel, tools, stationery, and office."

"I doubled my income the first year. I became my own boss and eliminated former stupid bosses."

"The major reward is in remaining alive technically."

"Getting paid for working on a rainy day and being able to play golf on a sunny day at midweek."

"I find being my own boss a tremendous reward. I call the shots and avoid jobs I don't like."

"There is a minimum of politics."

"A leisurely lifestyle at approximately the same income—I am retired from a major company. I only supplement my income."

Success stories further pinpoint what turns on engineering consultants.

"The corrosion of brazed joints in heater tubes for an 88,000-ton oil tank cost a shipbuilder a bundle of money. I traced the problem to a phosphorus bearing brazing alloy. To solve it, I switched to a silver brazing alloy. The client was extremely pleased and has since called me on other problems. My work also permitted some lawsuits to be settled out of court."

"The day after a young woman was killed and a young man was crippled for life in a motorcycle accident, I was called in to examine the new bike and discovered that a critical component, an aluminum die casting, was defective and broke for that reason, causing the accident. Subsequently the young man was awarded a multimillion judgment in court as the result of my findings."

Incomes Vary From "Peanuts" to Six Figures

When the discussion narrows down to precise numbers, the best one can expect—for several reasons—is a ballpark estimate based on examples of what others have done and are doing in terms of actual dollars (Table 1 serves this function); at the risk of being tedious, one must be reminded of the standard disclaimer, "It depends largely on how much you want to put into it."

Some of the information in Table 1 must be qualified: "1st Year" and "2nd Year" may be as long as 20 to 30 years ago; "now" figures are usually for the year 1982; and explanations were not obtained for how "net" was calculated. One may safely assume, however, that tax deductions were taken for such items as use of a home office, a car for business purposes, office equipment, technical books, etc. How much such benefits of being self-employed mean varies, of course. One consultant volunteered, "My tax accountant is able to get my net down to about 50% of gross."

TABLE 1
Consultants Report Earnings at Three Stages in Their Careers

First Year		Second Year		Now	
Gross	Net	Gross	Net	Gross	Net
$20,000	...	$30,000	...	$75,000	...
20,000	...	30,000	...	35,000	...
35,000	$29,500	125,000	$115,000	200,000	$145,000
...	15,000	...	30,000	...	75,000
...	15,000	...	25,000	retired	
30,000	5,000	...	...	130,000	65,000
80,000	...	96,000	...	118,000	...
15,400	10,800	16,200	12,900	17,600	10,500
30,000	20,000	30,000	20,000	160,000	85,000
25,000	15,000	...	...	21,000	12,000
50,000	...	60,000	...	tapering off to retire	
22,500	...	23,900	...	36,000	...
17,000	14,000	20,000	17,000	100,000	50,000
30,000–40,000	...	30,000–40,000	...	30,000–40,000	...
...	...	...	...	30,000	15,000–20,000
40,000	32,000	60,000	45,000	125,000	...
6,000	5,200	7,000	6,300	...	...
13,000	9,000	20,000	15,000	100,000	90,000
25,000	15,000	...	...	50,000	40,000
15,000	...	...	...	160,000–180,000	50 to 60% of gross
...	10,000	...	...	...	30,000
60,000	30,000	70,000	35,000	100,000	75,000
20,000	13,000	10,000	6,000	50,000	37,000
20,000 (half year)	11,000	43,000	30,000	60,000	43,000
19,000	10,000	25,000	15,000	...	...

Another thing—earnings tend to fluctuate. A consultant who did not report his earnings for this study explained, "They vary too widely to be meaningful." Another offered, "I may make as little as $10,000 in a year, or as much as $20,000; but I am retired, and for me consulting is more of a hobby than a business."

There is an approximate correlation between amount of time worked and amount of income realized. On the other hand, some part-timers seem to do as well as, and possibly better than, some full-timers.

"Since retiring, I consult about half-time. Gross and net would have to be converted to equivalent full-time employment to come up with an annual figure. As a consultant I earn $75,000 per year plus expenses, based on $500 per day and 15 days."

"In my first year—the last six months of 1982—I made $9,000. In the first six months of 1983 I made $14,000. I am retired. This money supplements my retirement and investment income."

"I work less than one day per week on the average and earn about $10,000 per year."

A semi-retired consultant "works about 1.5 days per week and grosses $30,000 to $40,000 per year."

To be absolutely realistic, one must remember that all consultants, both full-timers and part-timers, do not make out that well.

"My net earnings were negative the first year. I broke even the second year. My work was mildly profitable the third year."

"In my first nine months as a new consultant I earned $1,600 and had about $1,100 in unreimbursed expenses. I am now in my second year, and so far [the first six months of 1983] I have had an income of about $1,800 and unreimbursible expenses of about $4,000."

One consultant reports his earnings as "peanuts."

It is apparent in reviewing Table 1 that those cases are the exception. In general, prospects for a beginning consultant are attractive. Comments supplementing the data in the table bear this out.

"I used my net of $65,000 in 1982 to buy a computer for my office, landscape, and invest."

"On a gross of around $115,000 per year, my expenses run $40,000 to $55,000."

"When I started 20 years ago, I grossed $15,000 to $20,000 per year. My gross is now $160,000 to $180,000. I net 50 to 60% of gross."

Going rates for fees provide another way of gaging potential. Representative fees are reported in the two sample schedules that follow.

Sample Schedule of Fees #1

Regular per diem fee ....$600/day, plus travel expenses from home; mileage in car, 25¢/mile
Office work $70/hour
Laboratory work ....$40/hour, except scanning electron microscope, $100/hour
Preparation of steel and heat treat specifications and standards ....Ordinarily in office at regular rate
Use of proprietary steels and heat treatments ....Fee by agreement
Lecturing to technical groups ....Contract price
Prototype design reviews ....Regular per diem—usually two weeks at client's plant
Value engineering a product ....Regular per diem—usually one to two weeks at client's plant
In-house seminars on steel and heat treat selection for engineers ....Contract price—usually two days
Heat treat facility appraisal ....Regular per diem—usually one to two days per plant
Shop or field problems ....Regular per diem plus laboratory expenses—usually one or two days
Steel cost reduction ....Usually contract price or 10% of first year's direct savings
Expert witness ....Investigation—same fee as field problem Testifying—regular per diem
Product performance improvement .....Per diem with maximum figure

Sample Schedule of Fees #2

On-tap technical service ....Retainer equal to one or two days per month and 12-month minimum. Unlimited calls at any time
Product liability cases ....$60/hour for consultation, examination, study, travel, report preparation, etc.
Other consulting ....$80/hour giving testimony for depositions or in court
Expenses ....Transportation, lodging, meals, laboratory fees, communication expenses, etc. Cost plus 10% administrative charge
Photographs $1/print
Mileage ....30¢/mile
Retainer fee ....Retainer of $1,000 may be required with a written authorization to proceed with consultation, all of which is credited on first billing. Any unused portion is refunded
Billing ....Payment requested within 30 days from date of statement; $1^1/_2\%$ per month is added for late payment

Chapter 10

Summary of Basics and Qualifications

The Basics

Consultants usually work for industry, lawyers and insurance companies, and teach part-time.

Industry is a favorite field, with legal work, particularly serving as expert witnesses, coming up fast in second place. Teaching includes but is not limited to the classroom. Consultants also conduct or speak (for pay) at technical conferences and seminars.

On the average, fees for legal work run $75 to $100 per hour and $750 per day; those for industry, $50 to $75 per hour and $400 to $600 per day. In both instances, clients pay job-related expenses, and fees are either set by consultants or negotiated with clients.

Pay for teaching part-time at technical schools and colleges averages $20 to $30 per hour, sometimes less. These rates are fixed by the educational institution.

Pay for government-related work is about on a par with that for teaching, and rates are fixed by the client.

The Qualifications

You stand a good chance of making it as a consultant if you:

- Enjoy being an engineer.
- Do not mind putting in long hours, perhaps more than you did as a payroll employee, doing what it takes to be both an engineer and a small businessman.
- Seek more job satisfaction, more opportunities for personal and professional accomplishment.
- Like the feeling of independence and have a strong need to be useful and productive.
- Have a salable reputation as an expert, or at least an impressive resumé at the start.
- Are a registered professional engineer—it often helps, but isn't always necessary.
- Have a flair for oral and written communication.

- Are over 30, preferably middle-aged or beyond.
- Are a good listener and talker, problem oriented, well organized, quietly confident; tend to be a conformist in dress and manner; work well with people and are liked by people.
- Can sell even though you may not be too comfortable at it.
- Have a nest egg to pull you through a slow start.
- Are not particularly bothered by the pressures of full workloads and tight deadlines.
- Do not mind a fair amount of local and out-of-town travel each month.
- Are not too proud to ask for help if you find you need it on a job, rather than try to bluff it through.
- Subscribe to a code of ethics based on values like integrity, honesty, objectivity, professionalism.
- Represent and conduct yourself in a manner described as "having good public relations."
- Are willing to accept the risks of being a small businessman in exchange for the prospect of greater financial and other rewards, including such intangibles as greater job satisfaction and the joys of being one's own boss.

Part II

Setting Up and Establishing a Practice

Chapter 11

Advertising and Selling One's Services

Overview

A consultant has to advertise and he has to sell to get business, unless consulting is a hobby for him or he resigned or retired with a lifetime supply of clients waiting in the wings.

The first choice of consultants is to advertise and sell by letting one's work and reputation do the talking via referrals and word-of-mouth recommendations of former associates, friends, peers, and previous clients.

Other "advertising" techniques used range from writing articles and giving papers to writing resumés and advertising in the Yellow Pages.

Other "selling" techniques range from using friends and personal contacts to advantage to giving free engineering advice and distributing flyers and brochures.

How Consultants Advertise: Their Views

Since reputation is what sells the consultant, he appreciates the power of selling via advertising. However, he would prefer only the subtle word-of-mouth recommendations as his chosen method of advertising.

The viewpoint is expressed, "No form of advertising is very effective. By far the best is performance by the consultant, followed by word-of-mouth advertising by the client."

About a dozen other techniques are used by consultants to get their names before their client public. Not all are conventional forms of advertising. The most used, but variously rated, techniques for promotion of a consulting business include:

- Writing articles and papers for journals and technical magazines.
- Writing technical books.
- Presenting papers at seminars, conferences, and technical meetings.
- Attending seminars, conferences, and technical meetings as a way of becoming known to prospective clients.

- Joining technical societies and trade associations as a way of becoming known to prospective clients.
- Distributing resumés that document experience and accomplishment.
- Passing out business cards as a way of putting one's name before the client public.

Other techniques used to a lesser extent include these forms of advertising:

- Buying listings in directories of consultants.
- Advertising in technical magazines and journals.
- Advertising in the Yellow Pages.
- Distributing flyers and brochures.
- Distributing name-imprinted novelty items like ballpoint pens and book matches.

In addition to these advertising promotion/marketing techniques, others are cited:

For example, a consultant seeking international business makes himself known and works with U.S. government trade missions, consulate commercial officers, and chambers of commerce.

Other consultants recommended:

"Get in touch with past associates."

"Serve on local and national committees of technical societies and trade associations."

"I sent letters to key industry people acquainted with my work, asking them to recommend me as a consultant to the foundry and materials industries."

"Made myself available to three technical advisory services."

"Work with a consulting firm, but not as an employee."

Consultants have minds of their own and are not quick to jump on a bandwagon merely because others are headed that way. This trait, at least in part, explains why opinion on the effectiveness of a given advertising technique varies so radically. With the possible exception of "novelty items," which seem to be in universal disfavor, you can elicit a "for" or "against" for any of the forms of advertising mentioned in this discussion.

One consultant believes, "Novelty items, Yellow Pages, writing books, and ads in directories are the least effective . . ."

While another feels, "I have found the Yellow Pages the most effective. They have attracted more business for me than anything else."

One consultant states, "You must continue to be seen at technical meetings. This is the most effective way of advertising one's self."

While another counters, "My attendance at technical meetings has never resulted in a dollar's worth of business."

One consultant claims, "Advertising in general does not appear to be very effective."

While another reports, "I find advertising in technical magazines and journals my most effective way of getting business."

Despite the latter position, consultants in general seem to feel as uncomfortable with advertising as they are with direct selling. They suggest the feeling that advertising may not be proper or professional. Perhaps the ideal is stated by the consultant who declares, "I operate entirely on my established scientific reputation. I have never spent a dollar on advertising or sought a client."

Or, "All clients for whom I have consulted have approached me. I have been rather widely known in metallurgy and ceramics."

Or, "My unexpected clients come from word-of-mouth recommendations—friends or engineers who know my experience."

In the final analysis, however, consultants are pragmatic and recognize that the simon-pure approach may have practical limitations.

You hear, "In one respect being a successful consultant is like being a politician—both have to be well known. For me, writing articles and books is the most effective way of doing this."

Or, "By attending technical meetings you have the opportunity to impress others with your technical knowledge, wisdom, and common sense."

Or, "You can establish yourself as an expert in your field by giving technical papers at seminars and conferences . . ."

Or, "Publishing and lecturing, I reach large audiences and give them a chance to evaluate me."

Or, "Flyers and brochures help to establish confidence in my ability."

Selling: Personal Contacts, Friends, Flyers, Telephone, Free Engineering

Given a choice, many consultants would probably avoid all direct selling, preferring to have their work and reputations do their marketing for them. Given a choice of direct selling or one alternative, they would probably opt for selling through friendships and personal contacts. Supplying free engineering is a close second. Techniques such as flyers, brochures, and the telephone rate a distant third.

A consensus position is stated thusly: "Friendships and personal

62

contacts account for 80 per cent of my business . . . most of my
high-class jobs come from former associations with companies and
friends at one company . . . they are my first line of en-
deavor . . . personal contacts are all-important; you need to be
known . . . much of my new business comes via referrals from friends
and other clients . . . clients tend to hire consultants they know or
take the word of somebody they know and respect . . . friendships
and personal contacts rate very high, if not at the top; most people
looking for a consultant ask friends for recommendations . . . most
jobs are referrals from people who know my capabilities, experience, and character . . . almost all my work has come from previous technical contacts . . . technical and personal contacts are most
effective . . . personal contacts and past business acquaintances are
most important, no matter what degree of friendship existed . . . professional friendships and contacts are most important.
All of my consulting comes by word of mouth from contacts I made
throughout the years . . . friendships help you learn about prospective jobs, and personal contacts help you obtain them . . . most
people tend to be loyal to their friends and personal acquaintances . . . 50 per cent of what's required to get business is having
good friends and knowing company officers who have respect for
your work and will refer you to other clients . . . a personal recommendation is one of the few ways to get new clients . . . you
must be known in the field you consult in . . ."

There are some dissents on the effectiveness of friendships:
"Friends often expect free advice . . . friendships are helpful but
should not be depended upon—even your friends have to justify
business decisions to their superiors . . . friendships rate zero with
me. It's a tough world. Friendship doesn't count. Personal contacts
do, 100 per cent . . . I shy away from friendships; they don't mix
with business and dollars . . . personal contacts are rarely if ever
a source of business . . ."

Some qualifications are placed on friendships and personal contacts. "They are important when they are backed by a solid operating record; otherwise, they are of no value . . . they are important provided one has established a good technical ethical
reputation . . ."

Beginners are advised to take full advantage of these two sales
tools.

"I suspect it is difficult to get started without a large number of
personal contacts with previous employers . . . until one has established his reputation and has a stable of clients, one must depend upon friends and personal contacts. In fact, most consultants
have had their former employer as their first client . . . friends and
personal contacts are important in establishing initial contacts; your

expertise and conduct determine how well you obtain and hold clients."

Free engineering as a selling tool is analogous to gift-giving in the hope of getting business in return. The practice is common.

"I frequently give free advice in early discussions, but it is never in writing . . . just enough to demonstrate my expertise and to whet their appetites . . . taking part in seminars without being paid for it is a form of free engineering; this leads to calls from all over North America . . . I give free engineering, but only by telephone . . . I may throw out a sampling of ideas . . . I always give it on short telephone calls as a matter of general good will without consideration of future business . . . have often given free engineering without the hope of getting business. In certain cases, such as when the government has a problem, I consider it my duty as a professional . . . sometimes, like off-the-cuff stuff that requires little or no effort on my part—to create good will . . . to small companies in the hope of future business . . . yes, in moments of extreme weakness . . . in fact, I give too much . . . to government and friends, but never to get business . . . only on the first visit . . . it's difficult not to; many, many people like to pick your brains on the pretext of possibly hiring you . . ."

A couple of "No's" are worthy of note.

"As a rule I do not give free engineering, but on many occasions I have helped people without expecting anything in return . . ."

"No, but during the first meeting with a prospective client I do not hide information. I try to sell myself by being free and open with vital information."

Flyers, brochures, letters, and the telephone are believed to have particular value for the beginning consultant.

"I did this in the past," says a consultant. "Don't now. They're useful only to let people know you are in business . . ."

Another reports, "I sent brochures to 25 prospects and had affirmative responses from five. I followed up by letter and by telephone."

Says another, "Letters plus brochures open the door and can lead to calls for jobs."

Says another, "I sent out flyers and brochures when I entered consulting . . . they produced very little business."

Says another, "I tried flyers at the beginning. Nothing worked. Perhaps it was my flyer. However, I reject (mentally) flyers from others no matter how well they are done."

Says another, "Flyers, brochures, letters, the telephone have been fairly effective for me, but there is no substitute for performance on the job."

In other words, experience with these selling tools is mixed.

Chapter 12

Setting Up an Office

Overview

When an engineer decides to hang out his shingle as a consultant and starts to think about what he should do to set up his practice in proper fashion, he immediately encounters a variety of rather mundane but practical questions.

What are the views of established consultants on the need for an office?

If one elects to call a spare room at home the office, how about a separate business phone?

Should one go to the expense of having letterhead stationery printed?

How often are laboratory facilities required?

Need for an Office Inside or Outside the Home

Most consulting firms are one-man operations based in a home office. Among members of the panel, 81 per cent felt an office, either at home or outside the home, is necessary for both part-time and full-time consultants.

They say the consultant needs a place to work and a base of operations for contacts by mail, telephone, or personal calls by clients.

Typically, one room in the house provides space for a desk, files, and reference materials. It may also be equipped to hold meetings with clients.

Further, one needs an office to get status for business tax reductions from the IRS.

The minority sets up a couple of conditions. "The need for an office depends upon the type of consulting being carried out and the volume of business one has. If the operation is small, you can see clients in their offices."

Office Location, Separate Phone, Business Letterhead

All panelists have offices—90 per cent in the home. Of the remaining 10 per cent, one is in a laboratory owned by the consultant. Only one with a home office reported having "a room set aside exclusively for that purpose." One has an office at home and another office elsewhere—in a lab. The panel includes retirees who regard consulting as a hobby, part-timers, and full-timers.

Of those with home offices, 31 per cent have separate business phones, while 69 per cent get by with home telephones. Most have answering services or message recorders for their telephones.

About 90 per cent have letterhead stationery containing standard information like postal address and telephone number. Business envelopes to go along with the stationery are also common, as are business cards. In some instances, consultants have their own invoices printed.

Some comments:

"I have an office at home and a separate business phone, but only one phone number. It doesn't matter. I live alone. I have business letterheads and invoices printed for my home office. I also have an office with the legal firm that retains me year-round; I usually go there every day."

"I have an office at home and a separate phone. A tip for those with outside offices. The consulting literature warns of a strong urge to move the office home the last two or three years in one's consulting career. The dangers are you'll lose information sources rapidly, and you'll encounter what is known as the honeydew complex. Your wife is handy, and you'll ask her, 'honey, do this; honey, do that.' "

This consultant adds, "If you consult even part-time, you must have an office, a telephone, and some form of secretarial service."

"I used to work out of a bedroom," says another. "I now have a spacious 10- by 40-foot office and beautiful surroundings. Now I can spread out and have clients over."

"A letterhead is imperative, and you need the services of a secretary."

Laboratory Requirements, How the Work Is Handled

Most consultants (85 per cent) need the services of a laboratory on a regular basis, but only 37 per cent own labs or have the use of one, and only 12 per cent rent lab space. It's common practice for 89 per cent to farm out their laboratory work. In some instances, consultants are able to use facilities owned by clients.

Comments reveal a number of practices and tips.

"I subcontract all lab work. The commercial labs I use are certified by the Department of Commerce and meet NASA standards. They do competent, reliable work at reasonable cost to me."

"I prefer to rent lab time and instruments, then add the cost to my bill to the client."

"I own microscopes for routine work, but hire all preparation work. I also farm out work requiring the electron microscope, chemical analysis, etc."

"I have always owned a combined research lab and consulting office. I farm out work when requirements are beyond my facilities, or rent lab space and do it myself."

"So far I haven't had the need for a lab, but it is bound to happen before long. I will probably farm out the work and supervise it closely."

"I use four different labs for legal work. This way I can take advantage of best available facilities and get prompt, accurate work. Some large aerospace companies do commercial lab work."

"I usually need the services of a lab, but my clients have them. I supervise the work I give them."

"I own a lab, but it is operated by someone else."

"I pay, on an hourly basis, for work done in metallography, chemical analysis, fractography, etc."

Office Equipment, Personal Computers, Secretarial Services

Overview

A consultant advises beginning consultants to splurge in buying equipment. "You must gamble a bit at the start, reflecting where you will be in two years."

Facts indicate consultants in general do not skimp in outfitting their offices, at home and elsewhere. What an individual consultant may regard as "the minimum required" may well include a copier, word processor, and/or home computer.

About one in three now owns a computer or computer/word processor, and the total is bound to grow.

Three-fourths use some form of secretarial service to type letters and reports, answer the telephone, file, and take dictation.

An Office Equipment Starter Set

In a typical office you will find any or perhaps all of the following: a desk or two; work chairs, easy chairs, guest chairs; filing cabinets; typewriter—standard, electric, or electric correcting; check writer; computer, possibly including printer; drafting table and equipment; one or two office telephones; telephone answering machine (answering service is the alternative); electronic calculator; copier; tape recorder; adding machine; word processor or computer/word processor; pencil sharpener; tape transcriber; technical library, including handbooks, technical books, standards, journals, magazines; dictaphone; cabinet for office supplies; a table or two; radio; alarm clock; shelf space for awards and business memorabilia; binocular microscope; 35-mm camera; side table; slide-making equipment; a few scenic pictures; storage space; separate storage and work space; lamps; TV; hand tools for inspection purposes; hardness tester; dye penetrant tester; and ultrasonic tester.

As one would expect, answers to "What's the minimum office

equipment you could get by with?" are notable for their variety. A set of representative inventories of home offices follows:

"Telephone answering machine, typewriter, calculator with printout, filing space, desk, computer with printout, and tape recorder . . . I will add a copier when necessary . . . except for the computer, this is the minimum . . ."

"Desk, typewriter, files, copier, phone, and electronic calculator . . . this is the minimum . . ."

"Desk and chair, file cabinets, book cases, typewriter, technical books, pencil sharpener, electronic calculator, reference library, stationery . . . the above are sufficient and probably represent the minimum needed."

"Telephone, typewriter, desks, bookcase—all essential. I also have a computer/word processor that has become essential to me. I do not necessarily recommend it to others."

"Desk, chair, typewriter, hand computer, file cabinet, guest chairs, and lamp . . . which I believe to be the minimum for my type of office."

"Full secretarial typing, tape transcriber, and filing facilities . . . not much is required to get going."

"Typewriter, calculator, copier, dictaphone, tape recorder, file cabinet, office supplies cabinet, books and book shelves, table, desks, and chairs . . . the minimum."

"Electric typewriter, telephone answering machine, filing cabinets, bookcases, and basic furnishings . . . about the minimum."

"Minimum depends upon the local availability of a well-equipped metallurgical testing laboratory."

"Telephone, typewriter, calculator, tape recorder, files, desk, side table, extra chairs, book shelves, and fairly extensive personal library . . . about the minimum."

"Desk, filing cabinets, bookcases, typewriter, and phone are the minimum, but typing can be farmed out."

"Desk, table, chairs, filing cabinet, scenic pictures for the walls, a few chairs, a business phone . . . the minimum."

"Typewriter, photocopier, desk and chairs, lamps, file cabinets, calculator, bookcases, telephone answering machine . . . about the minimum."

"Filing cabinets and wife's typewriter."

One in Three Owns a Personal Computer

Three questions were asked:
"Have a personal computer?"

"Is one necessary?"

"Rent computer time?"

Results were as follows:

Thirty-three per cent own computers.

Eighty-nine per cent feel they are not essential to what they are doing at this time.

No one reported buying time.

Taking individual comments into consideration, the apparently high 89 per cent who feel computers aren't essential is misleading. A more realistic interpretation reads, "I would love to have one and intend to buy one as soon as I can justify the expense, or perhaps sooner."

A cross section of comments:

"I have use of a computer terminal for information retrieval, but I haven't felt the need for any other type."

"Nowadays a computer and terminal would be an absolute must," declares a retired consultant.

"I have three computers. I feel they are necessary but hardly justifiable—based on their cost alone. They have saved me from errors on repetitious items."

"I own a computer but have not learned how to use it . . . it's a good idea for handling accounts receivable."

"I bought a computer because it was needed for the work I was doing."

"I am just now learning to use my word processor as a computer, for financial records."

"Do not own one, but am planning to get one . . . not a real necessity, but almost one if one is to keep up with progress."

"As clients increase in number, buying a computer looks more desirable."

"Do not feel a computer is necessary for my consulting business. However, I have an Apple for my other business."

" . . . not essential, but I should probably have one."

"I don't have one, but I can see they have much value for a consultant."

"For me a computer is essential because I do all my own records, bookkeeping, writing, etc."

"I bought one to get the experience . . . one must know about computers these days . . . the actual need may be in the future."

Some minority views were expressed. One computer owner, for example, feels "it isn't necessary—a secretary would be as good." Also mentioned are electronic calculators and slide rules as "adequate alternatives for the work I am doing."

How the Office Work Is Handled

Who does the typing?

For 75 per cent, by a full- or part-time secretary, secretarial service, wife, daughter; or the "client provides secretarial services."

Who answers the phone?

For 60 per cent, an answering service, answering device, secretary, or wife. The remainder do it themselves.

How are letters and reports written?

For 75 per cent, in longhand and then typed by self or someone else.

The remainder dictate on tape or to a secretary or wife. Some clients provide such services.

A surprising 25 per cent prefer the do-it-yourself route to typing. One reports, "I recently acquired a word processor and do all letters and reports myself." Others use standard typewriters, electric typewriters, electric correcting typewriters, word processors, and computer/word processors.

Consultants display unusual ingenuity in working out arrangements for typing and other secretarial services.

"My wife types letters and invoices; my former secretary types my reports; and I generally write letters and reports in longhand."

"I rent an office from a typist who has a copier. She amounts to a part-time secretary for me."

"My secretary is an accountant by profession. She lives 50 miles away. We communicate by mail."

"For five years, typing and phone answering have been handled by a secretary in the building where I rent an office."

Telephone answering devices seem to be slight favorites over answering services, but a consultant cites a limitation of the former: "50 per cent of the people who call won't talk to a box." One advantage is that the machines can operate around the clock and seven days a week if desired. They are usually turned on when the consultant is out of his office. Answering service hours are typically limited to daytime business hours and five days per week.

Chapter 14

Largely Legal Matters

Overview

About one out of every three members of the panel is incorporated, often with wife or son or both as officers. At the time they took the step, they gained tax advantages relating to pension plans and insurance. Liability protection is not on a par with that of a corporation in industry. Insurance against liability is often recommended, but is far from universal practice.

Consultants with partners, silent or otherwise, look for such qualifications as compatibility, integrity, and reputation. Some wives are listed as partners. In one instance, "my wife and the good Lord."

Lawyers, accountants, tax consultants, and bankers may be needed on special occasions, such as at tax filing time or when cash flow dwindles and a loan is needed to keep the operation afloat.

Need to Incorporate, Family Members as Officers

About 34 per cent are incorporated.

Among those not incorporated, 45 per cent feel such action is necessary and about the same number say "it depends upon conditions," or "I intend to later on."

In a few instances, the corporations have wives, sons, and other family members as officers.

A sampling of views on the need to incorporate:

"After looking into the matter, I could see no benefit in it for me."

"Not for me as long as I can keep my retirement fringe benefits such as health insurance for me and my wife."

"It depends on the level of business. For me, the answer is 'No.' There is no reason to incorporate."

"I took advantage of tax benefits to improve my fringe benefits—such as building up a retirement fund along with medical and life insurance."

"It's probably desirable, but I would prefer not to."

"I carefully checked this out. While 50 per cent of all consultants are incorporated, my lawyer thought it offered only marginal protection for the cost. It might be required for a larger operation."

"You should incorporate. My son is president; my wife is vice president. I am listed as owner."

"Generally desirable, but not for me on my retirement basis."

"Not for me . . . others have found sufficient tax incentives."

"I have not. If I were younger, I would."

"My wife and oldest son are officers."

"Depends on the tax situation for the individual."

"Depends on the size of the business and number of people involved."

"Depends on circumstances . . ."

"Should as earnings justify it."

"Some types of work demand protection you can get only from incorporation."

"Worth considering. Prentice-Hall and other publishers sell tax manuals for tightly held corporations."

Partner, Silent Partner, What to Look For

Only 9 per cent have partners.

What to look for in a partner, they advise, is compatibility, competence, integrity, and a good reputation, along with knowledge and experience complementing that of the other partner or partners.

Other tips: " . . . have had a partner twice to augment and broaden skills to offer clients . . . partners should complement each other. They should not have identical backgrounds and competences."

Some of the benefits of having a partner are obtained through subcontracting jobs, in whole or part, to another consultant. "I sometimes employ another consultant if I need expertise in a related field . . . I have a friend to whom some jobs can be delegated. His background is similar to mine, but he has more experience in the automotive field. He also has a small machine shop in his garage, and he has industrial contacts that I don't."

Some prefer to go alone. "I intend to continue to work alone. I work on a part-time basis and only on jobs of interest to me. I do not depend upon consulting for a livelihood, and I do not want to be responsible for someone else's career and earning capacity."

Some (14 per cent) consultants have silent partners who may be "an associate who can be as active as advisable or required," or more likely, the partner is "my wife, who types reports, invoices, and letters to clients . . ." and, it is commonly reported, "she is not always so silent."

Other comments:

"No partner, but I refer inquiries to other consultants without remuneration. Perhaps they will return the favor."

"On certain types of jobs I work with other independent consultants—as in welding, coating, and water treatment."

Need for Lawyer, Accountant, Tax Consultant, Banker

Thirty-four per cent feel it is necessary to have a lawyer.

Beginners, in particular, it is felt, "need a lawyer when they are getting started—to answer questions about incorporation and liability."

The usual approach is to become acquainted with a lawyer and call on him when a problem arises or to review a contract, for example. In one instance, "My lawyer is secretary to the corporation, and in this role writes corporate resolutions and prepares the minutes of annual meetings."

In general, it is advisable to have a lawyer on tap "if your work is primarily technical and if you employ others."

Thirty-eight per cent feel it is necessary to have an accountant. Again, these services are regarded as essential for the beginner "for general guidance and at tax time."

Consultants usually feel that they or a secretary can handle normal bookkeeping chores—particularly for one-man operations, while two- or three-man operations normally have accountants or use accounting services.

Fifty per cent feel it is necessary to have a tax consultant, but the consultant may be the same person or service who handles other accounting jobs. Some hire CPA's.

Protection is not guaranteed. "I just had an IRS audit," reports a consultant, "and a tax consultant had helped me prepare the return." Another adds, "I had a tax consultant for five years and was audited. I submitted my 1982 return on my own." Another says, "I have taken the time to study IRS regulations carefully and do my own tax returns."

Twenty-six per cent feel it is necessary to have a banker or at least a bank.

Consultants often maintain separate business accounts, and having a bank contact to talk to is deemed desirable "if the business generates a fair amount of money or money must be borrowed from time to time when cash flow is slack to meet obligations."

Chapter 15

Some Financial Matters

Overview

More often than not, consultants give clients 30 days to pay, with no discount incentive. However, a penalty, typically $1^1/_2$ per cent, is added for slow pay after the account is, say, 30 to 60 days delinquent.

Several techniques are used to deal with slow pay, including telephoning, letter writing, and re-issuing of invoices. Some procedures are fairly detailed: "I call accounts receivable to see if the invoice has been received. Ask when payment is scheduled. Send invoice if one was not received. If nothing happens, send another within a week, and so on. As a final resort, I start to call contact I worked with at client's." Threats do not work, consultants advise, and add another deterrent. If you are not patient and get too aggressive, you may lose the client—if you value his business.

Slow pay and no pay are not universal problems. In ten years of practice, reports a consultant, "I have not had to use a lawyer or collection agency to get my money."

No Terms, Informal Terms, Highly Structured Terms Reported

"Net 30 days from date of invoice" seems to be the most common practice (about 40 per cent use it) among the dozen or more practices reported.

At one end of the spectrum, consultants "have no plan," or, "have had no need to date," or "have no official terms."

One gentleman confides, "I have never stated terms, but when I am asked, I say, 'net 30 days.' About two-thirds of my billings are handled by the organization [commercial testing laboratory] I work for. Their follow-through has been poor, but is improving."

Another approach in this category: "Terms are generally whatever the client desires. They usually pay in 30 to 60 days. I have no discounts or penalties."

In the next grouping are "payment at completion of services," or "cash on completion of work, or phase of the work," or "cash upon billing," or "payment on delivery of report," or "invoice payable on presentation at end of month."

A consultant who specifies "due and payable on receipt of invoice" does not offer any discounts, but does not demand penalties for late pay. "Of 110 invoices issued last year," he says, "the average delay was 44 days."

A consultant specializing in legal work requests "net 10 to 20 days after work completed . . . advance payment of up to $1,000 for plaintiff attorneys or a $500 advance for good-pay plaintiff attorneys, or $1,000 on first job for an attorney or, on occasion, for an unknown defense attorney. I refuse to work for some plaintiff attorneys. They never pay."

Another approach: "I bill for services at the end of each month and expect payment during the following month."

There are some variations on 30-day terms, as "30 days plus $1\frac{1}{2}$ per cent per month for late payment," or "net 30," or "30 days from first of month following billing date," or "30 days from date of invoice."

How Consultants Deal With Slow Pay

For some, "There have been no problems with this," or "I have not had much of a problem."

If there is a problem, everyone seems to have countermeasures.

"If I need to, I follow up monthly, and go higher in the organization if necessary."

"I rarely encounter this. If so, I usually send a duplicate statement, asking for the client's assistance in obtaining or expediting payment."

"No problem. If pay is slow, I ask my direct contact at the client's place if he received the invoice, and this usually does the trick. Even this is rarely necessary."

One consultant adds the thought, "Payment usually takes 60 to 90 days. You may not want to enforce terms in the hope of getting repeat business from the errant client."

Others who think along the same lines say, "I deal with slow pay by crying," or "wait in silence," or "deal with it patiently."

The majority is shorter on patience. Techniques recommended include: . . . "re-invoice and telephone . . . keep telephoning . . . make calls and write letters . . . after 60 days, re-issue the statement. . . phone the general manager or project manager . . . telephone and ask for an explanation of the delay." A consultant adds, "I pursue the matter by telephone call and letter billings, according to my judgment of the financial capability of the client."

Many procedures are available.

"No magic formula. If you expect trouble, ask for prepayment. Same thing on repeat business from a slow-pay client."

"Ask for payment in 30 days. Follow up in 45 days. The next time around you may want to ask for money up front."

"Persistent reminders, written and oral. After that, a demand notice [statement of claim] from my lawyer."

"Both personal and written appeals. On occasion, legal action is required. Collection agencies are sometimes effective. Threats seldom if ever work."

"After 30 days, I send a second notice. After 50 days, I call the person who hired me."

"Most clients pay within 30 days. I re-bill each 30 days thereafter. In nine years only one company did not pay. Went bankrupt and could not be located . . ."

"Make sure they got the invoice. Then I contact my personal contact at the client's place. Often there is only a minor glitch. Next I call accounts payable. Be persistent. I am considering small claims court in one instance."

Adding a penalty is a popular technique.

"I believe the best procedure is to specify a cumulative interest charge in the invoice that will be added within a specified time—usually 30 to 60 days."

"In some cases I add 5 to 10 per cent if payment isn't within 60 days. However, my experience has been quite good—clients normally pay promptly."

"I send monthly invoices repeatedly . . . add 1.5 per cent penalty to bill each time."

At times, a clever response pays off. "I have had only one problem with slow pay so far—an attorney," reports a consultant. "I called and asked him if I could hire him to collect. I received a check promptly in the mail."

Only about 20 per cent have ever hired a lawyer or collection agency to pursue a truant client.

Among those who have not had the problem, you hear, "No, and I would not hire a lawyer or a collection agency . . . I have not found it necessary in ten years . . . no, but I am thinking about it . . . no, I work mostly for attorneys, but insurance companies are often slow to pay . . . no, but if it is a legitimate invoice, is undisputed, and I do not care about repeat business, I probably would go after him."

Others report, "I have used lawyers or collection agencies a number of times . . . I lose about one payment per year—a total of $500 to $800 . . . only once. A lawyer. I wrote it off . . . twice in 6^1/$_2$ years, for a total of less than $2,000. I collected 75 per cent of it . . . I have an attorney friend who will compose a demand notice for a modest fee . . ."

Chapter 16

Up-Front Money and Contract Language

Overview

Opinion on the propriety of asking for money up front is about evenly divided between two groups. For one, preferred practice seems to be, "pay it out of your own pocket and bill the client later." For the other, it is considered proper to ask for prepayment—usually partial—under certain circumstances, as when expensive travel is required.

Questions of ethics often arise when performance or contingency contracts are discussed. It is not unusual to hear, "I do not do this type of consulting."

In writing contracts, brevity and simplicity are favored in principle. The resolve may be diluted by the engineer's penchant for detail and specificity.

The Up-Front Money, Advance Question

"Up-front money" and "advances" are regarded as different terms. The latter is confined typically to expenses related to travel, while the jurisdiction of the former includes expenses in general and even partial payment of the consultant's fee. About 40 per cent ask for money up front, for example, to cover out-of-pocket expenses like lab work, or travel, or to provide security where the client—particularly one who is a lawyer—is an unknown or has a reputation for slow pay.

As expected, a sizable number of consultants do not ask for up-front money or advances "because it does not strike me as being professional," or where "it is understood my time will be paid for even if the job is terminated early," or "it is not necessary for the people I work with."

As an alternative, a client may be asked for a blank purchase order up front, or a purchase order for a commercial lab, for example, doing work for the consultant.

Comments reveal up-front money is requested under a variety of circumstances.

"Yes, with some clients I ask for money up front."

"I haven't yet, but I would on a contract of two weeks or more and up-front expenses were heavy, such as for international travel."

"I have not had to. However, increasing experience with lawyers as clients indicates it is a good idea and should be done in such cases."

"It is good practice to ask for a retainer when working for a plaintiff in a court case."

"Sometimes I ask for money up front if the client is unknown or has a dubious reputation."

"In most legal cases, yes."

"Yes, with the few clients that have a reputation for delinquency."

"Not so far, but I am in the process of changing this. I anticipate significant out-of-pocket expenses on a job."

"I ask for $300 to $400 to pay laboratory and exemplar costs . . . to be subtracted from the first billing."

"Sometimes, when early expenses are high."

"I do if I am not sure of a client's credit rating."

Asking for an advance for transportation costs seems to be frowned upon less than a request for up-front money. However, some consultants still take a hard line. They say, "Clients occasionally offer to pay for airline and travel expenses up front, but we usually pay these expenses and bill the client . . . or, for air and other travel I pay my own way and bill later . . . or, standard practice is to bill after the trip. I feel this is much more professional than asking for an advance. I present an itemized account of my expenses, using my own expense form."

Prevailing opinion runs like this:

"I ask for an advance for air fare to South America or to Sweden to attend a technical conference—this is normal practice."

"I request an advance only when the client asks me to visit him."

"I ask for an advance only when travel costs exceed $250; then I request 50 per cent of the sum as an advance."

"I ask for a travel advance when the client is new to me or I feel he may not pay on time—but only for long-distance travel expenses."

"Some clients provide plane tickets."

"Sometimes the client will offer to pay expenses or send a company plane. If so, I accept. If not, I pay my own way and bill later."

"I ask for a travel advance only for out-of-the-country consulting. Usually request a round-trip ticket."

Performance/Contingency Contracts: Usually, "No"

Only about one in five accepts or has had any experience with this type of contract. The majority position is strong.

"I would not accept such an arrangement . . . consider it unethical."

"No—especially with an attorney."

"I haven't encountered such contracts, but I doubt if I would accept."

"They are unethical in product liability cases."

"I do not do this kind of consulting."

"I get my fee only if the lawyer wins his case? No dice!"

"I take on work only if I feel I can handle it. Otherwise, I recommend another consultant."

The minority view:

"Fine, but it can be sticky if the contract is not carefully drawn."

"Have not faced this, but would not hesitate to discuss the possibility in my field."

"I have had only one job where payment was tied to satisfactory results. However, I was positive of success."

"I have never had any problems with them."

"Might consider if I had unusual interest in the problem."

"I once worked with a bonus clause."

Contract Language: What to Include and Exclude

"The question of personal liability should be covered . . . "

"Be brief. Most of the time my contracts run one to two pages. The longer the contract, the more you are in lawyer's land."

"You probably need a lawyer if you need a contract."

Those comments indicate the range of opinion in this area. Many tips are advanced.

"Include what is to be done and what you will be paid."

"Include protection for yourself and ensure payment. Try to exclude anything that ties you to a specific end result, unless it is attainable and results are measurable."

"I prefer a simple letter. Then common law applies."

"You should include extent of liability, length of time required for job, fee or hourly rate, requirements for report or reports, and an outline of the tasks to be performed."

"Always include a clear and brief milestone chart . . . what will be accomplished and when."

"Include who is responsible for payments, when payments will be made, and extent of payments. Include, as precisely as possible, when work is considered done and delivery completed."

"Include fee and an expected payment schedule, generally a description of the work (but not too restrictive). Exclude insurance and 'hold harmless' clauses."

"My contracts usually include secrecy agreement not to work for a competitive company during the duration of the contract or longer."

"Most particularly include, 'This contract constitutes the entire agreement between the consultant and client.'"

". . . don't encumber the contract with too much detail, as in many government contracts."

"In most cases my contracts are verbal . . . only occasionally is a brief letter involved."

"Define extent of services clearly. Otherwise, you'll cover a whole range of problem areas."

Chapter 17

Dealing With Clients, Client Personnel

Overview

Client personnel that must be dealt with directly often feel threatened because a consultant has been called on the case, requiring him to practice diplomacy.

Among the many techniques reported are the following:

1. Enter the group (client personnel) with modesty. Express your admiration for what they do, and make it known that the only thing you hope to bring to the situation is a fresh viewpoint—because of your experience and specialization.
2. Go into a listening phase. They know much more than you do about the whole layout; ask them to educate you on the background of the problem; ask for their opinions since they have given the problem much more thought than you have up to this point.
3. Step forth with your suggestions, and avoid contradicting the in-house experts.
4. End with a summary of what you expect to do.

Most consultants feel that, as professionals, businesslike rather than personal relationships with clients should be maintained. "We may meet for dinner to discuss the problem, or go to lunch; relationships are pleasant and friendly, but not social."

The Threatened Employee Is a Common Problem

Seventy-six per cent must deal with client personnel who fear for their job security because a consultant has been summoned to solve a problem that has them stumped. The situation is encountered in varying degrees—from "rarely" and "seldom/sometimes/occasionally" to "often," "quite often," and "always." It happens only in work for industry. There is no confrontation in this sense in the classroom or in the courtroom.

The person who hired you—the client, the boss—is not involved.

The challenge is in coping with his employees, people you must deal with directly on a daily basis in working on a problem. You need their cooperation and good will. The way to get them can be summed up in a single word: diplomacy.

Consultants are articulate on the subject and have a wealth of practical wisdom to offer.

"It is often better to induce ideas rather than force them . . ."

"Sometimes you have to work around individuals or groups who feel threatened."

"They are on the defensive. You ask them to tell you how the problem developed—to get their views . . ."

"I am sympathetic to an attitude of resistance among company engineers; my recommendations often concur with what they have been telling their management. I try to sell them on the idea to use me to get their objectives into my final recommendations."

"Compliment them and ask for their help."

"I try to deal at the highest possible level—you are not threatening *him*. With the others, I try to be as friendly as possible, explain what I am doing, and cultivate cooperation."

"I praise the people I work with, but when I think they are wrong, I suggest alternatives."

"Rarely have the problem. If I do, I deal with it by attempting to gain their confidence and make them realize I am concerned only with technical issues."

"I get into friendly discussions and forward the thought that what ails the company concerns everyone. Sometimes a compliment to the person or group that feels threatened (particularly in their presence) helps a lot."

"I run into this, but ignore it and go about the job objectively. When they see I am pointed at the problem and not them, they usually come around—although it may take them several hours to realize I am there to help them."

"Find some way to give them recognition or credit."

"I have them work with me in defining the problem, and I keep them informed, on a step-by-step basis, of what test results indicate, and tell them what my final recommendations will be."

"I am careful to avoid any impression that I feel superior. I discuss, rather than orate, for example."

"With the permission of company management, I take these people into full confidence, explaining why I am there, what I expect to accomplish, and how it will benefit the company. If management won't allow this, it would be better to decline the job."

"I take command as diplomatically as possible by explaining I have been on both sides and understand their reaction to me; that

I am there as a problem solver, not a hatchet man; and I need their cooperation in providing the facts and data required. The alternative is: if I fail, more consultants may follow me."

"In a very diplomatic manner I point out that the problem is an unusual one and that I require their help in solving it."

"There is some of this on every job . . . among other things, I try to size up the problem and its effect on different people I must deal with. I do not blame anyone for failures. I try to protect as much as possible. This works most of the time."

"Firmly and politely, I point out what I am hired to do. I solicit cooperation so we can work together to find out what is wrong and determine how to correct it. Avoid any discussion of who is responsible."

"If I run into this, I let the person know I am not interested in his job. I am working only to benefit him and his employer."

"I try to make sure they can claim credit for success."

"If resistance can be anticipated, one can conduct himself in a way not to offend anyone. I try to get these people involved in helping me, so they can become part of the effort."

"I encounter the problem in the majority of assignments. A very humble and respectful approach is necessary. Selling yourself to these people is an important part of consulting."

"Be as friendly as possible, and above all be pleasant. In my experience, the problem is much less frequent than one would expect."

"I make sure my primary contacts are with management. They do not feel threatened by my presence."

"I tell the person who seems disturbed how good he is and emphasize that my job is mainly to ask the right questions."

"Get them on the team and somehow feed them information to make them look like heroes to their bosses. But you must control the work and the program. Acknowledge them and praise them whenever possible to their management."

"Do everything possible to indicate you are a member of their team. Avoid criticism. Feed each of them ideas—individually and confidentially—that they can use."

Personal or Businesslike Relationship With Clients?

Sixty per cent favor a businesslike relationship with clients, while about 30 per cent report their relationships are a mixture of "personal" and "businesslike." Only about 10 per cent work strictly on a personal basis. Plausible explanations are offered for each position. You hear:

"It is a business transaction. If one goes beyond this, the client might suspect you do it with others and do not maintain the confidentiality you should."

" 'Personal' runs into complications and telephone consulting."

"I am businesslike, but this does not prevent me from being friendly."

"When the relationship is businesslike, you maintain mental respect."

"It is better to be very impersonal, so that your recommendations are based entirely on facts, because they [the recommendations] may be at odds with what others think."

"Businesslike, unless the client is a personal friend."

"Consulting is a business."

"Getting too friendly with top brass makes it harder to deal with lower echelon troops in getting the job done."

"You must be professional."

"My relationship is one of furnishing technical expertise . . . "

Among those who favor a personal relationship:

"In working with client personnel, this is the only way to find out what is going on from the inside. Also, most businesslike relationships have a way of turning into personal ones. They [client personnel] appreciate you helping them look good in the eyes of their management."

"This is the way I operate best, and it allows for a more pleasant relationship."

"I maintain a personal relationship with longer term industrial accounts and with some law firms, because much of my work comes from referrals. I also work closely with testing laboratories."

"I usually prefer the personal approach because it is informal, and I feel more comfortable that way."

"Personal, but always businesslike."

The mid-position group includes those who maintain a personal relationship with Client A and a businesslike relationship with Client B. Many times, a relationship that starts out businesslike mellows into a personal one.

"In some cases I already know the client and previously had a close relationship with him—one which continues when I work for him. In other cases, I maintain a businesslike relationship from beginning to end."

"Generally they start out businesslike; then they frequently become quite friendly."

"Business relationships usually become personal after a few meetings."

"Businesslike with most clients; semi-personal with some."

Chapter 18

More on Relationships With Clients

Overview

"In my one-man show," confides a consultant, "it is important to develop a personal, day-to-day relationship with clients and client personnel, as long as it is compatible and does not compromise anything."

In calling on clients, most consultants opt for a conservative, professional appearance. However, some ascribe to an affluent style, "if you don't do it for display."

Developing Professional, Yet Friendly Relations

Even though consultants feel it is essential to be businesslike in a professional sense, they recognize that consulting is also a business; and in this sense it is advisable to develop and maintain good customer relations. In this context, about 75 per cent feel they should be on friendly terms with clients and client personnel.

There is little disagreement over the importance of this aspect of operating a business. Many believe "it is as important as the technical aspects of a project." Any disagreement centers on semantics—the exact nature of the desired relationship and what it should be called.

You hear:

"Depends on how you define personal relationship. A good working relationship is very important, but it doesn't have to be personal in the special sense of the word."

"If you mean away from the office, personal relationships with clients are not important to the consultant. At the office, it is well to adopt the degree of formality or informality comfortable for you. But don't pretend."

"I would emphasize friendly cooperation rather than personal relationships."

"Business should be on a friendly basis without being pushy."

"Friendly, but not buddy-buddy."

"The important thing is a cordial, friendly relationship based on giving good technical advice."

"The relationship with client personnel should be comparable to that between two employees."

"It is important to develop a relationship with the client so he trusts you and values your judgment and competence. This may or may not be achieved through personal relationships."

Other views on the same position:

"I always want to be on a first-name basis and consider the client a friend."

"Friendly, personal relationships are quite important. Frequently they will lead to recommendations to other divisions within a large corporation, as well as to other companies."

"I have known most of my clients for many years."

"Mutual respect over the years leads to an open door."

"Few clients become personal friends, but I am friendly to them always."

"Developing a relationship is of intermediate importance for the short term . . . necessary for the long term."

"I feel there should be little or no personal relationship. Of course, this does not mean a lack of friendliness, which I take to be different."

Dissenters say:

"I prefer to keep it on a businesslike plane."

"Not important . . . results are what count."

"If you do a good job for them, the personal relationships will take care of themselves."

Should You Go First Class in Visits to Clients?

Should you project a prosperous, even affluent, look when you are out on a call?

Or should you maintain a low profile?

Both views have advocates, but the consensus takes a middle position " . . . not necessarily affluent, but not cheap either . . . I have never been impressed by an affluent appearance; however, I feel good grooming and a professional manner are important . . . I fly coach and stay at moderately priced hotels, but I dress well and try to act in a professional manner . . . I try to economize on expenses, but follow the golden rule: never cheat yourself . . . be reasonable, but dress well; stay at good hotels and eat well . . . I do as I would if I traveled at my own expense . . . I fly coach and spend the client's money as I would my own. I stay at first-class hotels, but not ultra . . . image is important, but excessive or phony affluence or economy should be avoided . . . moderation is the key . . . good hotel, fly first class . . . I follow my normal lifestyle, which tends to be economical, but not parsimonious . . . I travel in the same style as I did as a corporate employee—no hotel rooms costing over $120 per night, fly tourist class . . . use common sense . . . live normally . . . do not splurge or scrimp . . ."

On the other side, some believe affluence " . . . probably helps . . . you must maintain an air of some affluence . . . clients respect your ability to handle yourself well when you travel . . . I generally go first class . . ."

The "against" contingent is larger.

"When I travel, I live as I would on a personal trip. Who am I trying to kid? I am not affluent."

"A comfortable lifestyle with a degree of economy. Fly coach, use compact car, standard motels."

"Affluence leaves an unfavorable impression on some clients."

"Try to be unobtrusive and reasonably economical."

"Clients appreciate when you economize, especially a corporate client involved in technology."

"I have found no need to puff."

"I economize. My Scottish ancestry, no doubt."

"Have always been thrifty . . . by Presbyterian predestination."

Chapter 19

Meal Meetings, Entertaining Clients

Overview

Meeting the client outside the confines of an office or plant isn't necessary, but can be helpful. It provides the opportunity to get better acquainted or to meet with key people you couldn't see otherwise.

That's one of the popular positions on the necessity for and productivity of meal meetings with clients and client personnel.

Any entertaining tends to be infrequent and on the modest side—mainly lunch, dinner, and perhaps a drink or two at these times. More ambitious socializing—golf, a play, a football game—is much less common.

Meal Meetings: Not Always Necessary, But Usually Productive

"They aren't necessary and most of them are ineffective."

"They aren't necessary but they can be very productive, such as where there is a need to get acquainted and time is limited. They are not as effective as an office meeting, though."

"They are necessary. Nearly all my contracts have been initiated at luncheon meetings. I do not entertain."

Of the three positions, the last two are the most prevalent. First, those who think meal meetings are necessary and productive.

"They are a pleasant and simple way of fitting a meeting into a busy man's schedule."

"Visual contact and personal contact are your most important sales tools. I always try to directly interact with my client, especially a new client."

"Meal meetings are very often necessary. They can be important to the client because he or she gains valuable information and saves time."

"Can be very helpful in getting a more complete background on problems and personnel."

"Time is very important in most situations. Meal meetings often conserve time; in addition, the really key individuals attend such meetings and are more likely than they would be otherwise to let their hair down in the relaxed atmosphere of a meal."

"I have accomplished much with top management at meal meetings."

"Helps people get acquainted in a pleasant atmosphere, along with having a business discussion."

"A great way to get to know people; the more casual, the better."

"With client personnel, meal meetings seem to take place automatically, and are often productive, especially in dispelling fears for job security."

The "not necessary, but" group is about the same size as the preceding one.

"Not necessary, but occasionally can be productive."

"Not necessary, but I go along with them if time is important. Sometimes all the people you wish to meet can only be free at lunch, or dinner, or even breakfast."

"On occasion, when it appears you can get a substantive discussion without interruption."

"Only when this is most convenient for both parties."

"Not necessary, but a lunch to discuss progress or to meet other company personnel is acceptable."

"Not necessary, but sometimes helpful to get a relaxed discussion of the problem."

"Not necessary, but may be helpful in getting to know people better."

On the negative side, a position like this is taken:

"I never do it, and I would be embarrassed by the thought it may be considered as politicking."

To Entertain, Not to Entertain

Curiously, about 70 per cent feel it is not necessary to entertain present and prospective clients, while about the same number do entertain on occasion.

The reasoning of those who feel it is necessary:

"I do this with established clients only. I do not try to buy my way into a consulting relationship with prospects."

"With regular clients, yes; but don't overdo it. With prospects, no."

"I entertain strictly in a business way, such as lunch . . . but no special entertainment."

"Nothing elaborate or expensive."

"Only occasionally when circumstances dictate, but not as a regular activity."

"I try to entertain occasionally, with larger clients. However, my experience is that clients are so enthused, they want to do the entertaining."

Generally, the "No" group bases its judgment largely on the propriety of entertaining, not the effectiveness of the activity for business purposes.

"Absolutely not" is the common statement. You also hear:

"I don't do it, but it certainly doesn't hurt."

"In my type of consulting, this isn't necessary. In fact, I have restricted my activity to avoid situations where entertaining is necessary."

"I don't entertain, period. Sometimes clients entertain me."

How Clients Are Entertained

More often than not, it's lunch or dinner and a ceremonial drink or two.

"Occasionally" and "seldom" seem to be the two common frequencies for entertaining, as " . . . when a successful operation is concluded . . . or when the client happens to be visiting in my area . . . or when it is important to get to know people better . . ."

Beyond the standard meal and a drink, clients are sometimes treated to " . . . an occasional football game . . . an evening at my home—as an alternative to night clubs and the like . . . plays, sports events, dances . . . a game of golf."

A few go further and send small gifts, such as a fruit cake, at Christmas. One sometimes provides transportation for clients "when they are in the area."

Chapter 20

Keeping Daily Records, Keeping Books

Overview

Note taking and bookkeeping.

Both very ordinary activities have particular significance for the consultant.

Daily events in the office and in the field vital to the conduct of business must be recorded and preserved in detailed and accurate notes, then kept readily available for later recall.

Records accounting for expenses and income must be maintained in a manner acceptable to the IRS and other government taxing bodies.

Daily Record: Calls From Clients, Other Business Events

Thursday morning. A client calls and outlines a new problem. You suggest a couple of preliminary ideas and eventually agree on an appointment for next Monday, 2 p.m., his plant.

The daily mail brings a bill for $218 from a commercial lab doing some radiographic work for a Job X, plus your final report on Job X, typed by your secretarial service from the longhand report you wrote on Friday last.

A prospective client visits your office at 11:18 a.m. and presents a problem you may not be qualified to handle alone. You intend to confer with a colleague on the possibility of a joint project. The prospect wants an answer, preferably by phone, no later than a week from next Monday.

How are such events recorded for reference at some later time?

To summon up precise details of the telephone meeting with the client?

Or the time of the visit to the client on Monday?

Or the date of the bill from the commercial lab so you will be sure to pay before it is overdue?

Or when the typed version of the final report became available for sending to the client?

Or to jog your memory on why you want to confer with a colleague before taking on the new assignment proposed by the visitor to your office?

The simplest and perhaps most common technique for recording such ordinary but essential intelligence is to take careful notes during or immediately after the happening—on writing tablets, stenographic pads, slips of paper, or whatever you can put your hands on at the moment. Curiously, there is no report of the tape recorder being used for this purpose, and perhaps it is just as well. Several disadvantages can be called up. In attempting to record a dialog, some people may tend to freeze or even regard the procedure as being improper in some manner. In addition, a step would be tacked onto the record-keeping procedure. Information on tapes would have to be transferred to a written record. Also, recording quality tends to vary; some parts of a tape may be garbled or even lost.

Consultants have adopted an array of procedures and methods for keeping records of daily events for later reference. Logbooks are popular as temporary or permanent repositories for such business information, as are an assortment of journals, diaries, calendars, tablets, and just plain note paper.

A number of procedures and practices have been developed:

"A log book is essential, especially for making out billings."

"I make notes on a yearly calendar. Day by day I write a summary of events, including the time each one happened, and I keep track of the time spent on work for each client."

"I set up a card file for each client. Take notes on $8^{1}/_{2}$ by 11 paper and place in appropriate client files. I augment these records with a field book and a diary."

"A logbook is a must. I have a separate book for each client."

"No logbook. I use a large desk calendar to keep a record of appointments."

"Always maintain these records in a logbook and daybook. Transactions are recorded in the order in which they take place."

"Keep a detailed, day-by-day log . . . needed for billing."

"Put my notes in file folders marked with appropriate headings."

"Always make entries in a stenographic pad near the telephone and later transfer the information to a file."

"Keep a telephone logbook and a daily log of time spent on various jobs."

"Have a separate notebook for important telephone calls and a leather-bound logbook for work performed and chargeable time, as well as industrial job titles and telephone numbers. A separate notebook is used on the job—pages may be later removed from it for inclusion in files."

"When a job is completed, a copy of the final report to the client is placed in the files."

"Keeping adequate daily records is a key point. I log all calls, expenses, reports, purchases, statements, payments, etc. It is a shortened diary to provide a chronological listing of my business events."

"Log is kept on a tablet or note paper and later transferred to the client's file."

"I have a phone call logbook that makes a carbon copy, keep a company and client card index, and make office memos regarding personal calls on clients."

"Logbook is especially important on a project . . . not only a daily record but also the exact time of events in a shop or lab or the occurrence of a problem."

"I have established an activities log for each client and a general log for other happenings."

"I maintain a daily account or diary, showing appointments, priorities for items to be done, hours worked, and expenses. This is in addition to an appointment calendar I carry with me."

"I take a datebook and expense book when I travel."

"Keep log on all contacts with clients, whether telephone calls or personal discussions. Logs are dated."

"I keep a daily log by the hour, which is valuable for IRS purposes. The more detail, the better. This is more information than your employer required."

Next Step: Bookkeeping System, If Any

How do consultants keep records of income and expenses?

The formality or informality of the system used seems to hinge largely on volume of business. A bookkeeping system may entail nothing more than putting hastily scribbled notes or other daily records into the appropriate client folder at the end of the day or some other time.

One consultant, for example, uses his daily diary as his final record and reports it is sufficient to "convince IRS that consulting is more than a hobby with me."

Another keeps a "log of expenses and income on a day-by-day basis."

Another writes up "detailed record slips for each job, on a weekly basis."

Another records, in a ledger, "hours worked for each client, along with associated expenses and income."

Still another keeps a "full set of books, using the double entry system. Information is taken from daily records of time and expenses, kept in a diary."

There are many variations of these practices.

"Use checkbook and credit card receipts."

"Keep track of billings and those paid; use receipts for expenses; keep record of time spent on each job."

"I keep a bound book, more or less a diary, into which all activities are entered."

"All out-of-pocket travel expenses are recorded in a travel expense book; others charged on American Express or Master Charge. Outside lab work is recorded in log of job."

"Use checkbook, charge card receipts, other receipts."

"Each statement I prepare contains itemized fees based on hours and days of service, plus a separate listing of expenses. Copies of statements are filed and used in tax preparation."

"Use invoices for records of income and company checks for expenses, unless a credit card is used."

"My books are a notebook and small computer."

"Being incorporated, I must keep books for the IRS, state, county, city, etc."

"I keep an expense register, a sales register, and a cash received register."

Some consultants do their own bookkeeping or have their secretaries or wives do it. Others turn to full- or part-time accounting services, accountants, or tax consultants.

Chapter 21

First Call: Who Pays? Retainers? Minimum Job?

Overview

The client, you would guess, probably pays for the first call. Not so. He pays—but not always. Some consultants think the initial visit should be a freebie if their primary intent is to do a selling job. However, if they take the trip at the client's behest, the latter, the thinking runs, should pick up the tab.

The reason for requesting a retainer may be, for example: the consultant regards it as security because he has a hunch or well-founded suspicion that the client will be slow in paying up. Or a hefty sum of expense money will be needed up front.

"Half day" is a common minimum for a job. As in the case of "who pays on the first call?" there is an opposite view. Some consultants are willing to take on a job of any size.

Consultant or Client: Who Pays on the First Call?

Forty-three per cent say, "The client."

Twenty-three per cent say, "I do not charge for this."

Thirty-three per cent say, "It depends."

Views and comments on the first position: ". . . most of the time the meter runs on the first call . . . the client pays, always . . . usually the company requesting my services . . . the client, if he calls me . . . when a potential client asks you to come to discuss the problem, he pays . . . the client basically pays for all time, but I do not charge for short (5- to 15-minute) telephone consultation . . . if he is in trouble and needs help, he is willing to pay. If he does not have a problem and just wants to pick my brain, I am not interested . . . normally, the client. If a job does not develop, I absorb the time . . ."

Among those who do not charge the first time:

"First call is by a lawyer asking for help. I usually do not charge for this brief introductory time."

"The business agreement is made on the first call—at no charge to the client."

"I normally do not charge for the first call or visit since this normally covers a description of the problem, and I do not do any work on it at this time."

"First visit is free. I pay. But the purpose is to set an agreement, not consult—except to give a sample to clinch the job."

The "it depends" group:

"I often charge the client, but not always. I keep track of the time, but decide later."

"Depends on circumstances. Usually there is no charge for a phone call from the client and a follow-through letter from me. But I charge on the follow-up visit."

"Depends on the agreement with the client."

"Varies with the amount and the importance of the service performed on the first call. I try to leave an opening to withdraw at this time. The client has the same privilege."

"If the call is in my home town and the visit is exploratory, I do not charge. If the visit is in town or otherwise follows one or two telephone calls indicating we will discuss business, they pay."

"The client . . . unless I am trying to promote business and have invited myself for a sales call."

"If it is a get-acquainted call, I absorb the cost. If the client wants to discuss possible work, I pay. At the proposal stage, I may split the cost. Once there is a contract, he pays."

A tip: "Beware of companies that even suggest a gratuitous first call. Class companies do not operate in this fashion."

When to Ask for a Retainer, and How Much

At least nine out of ten ask for retainers. They give a grab bag of reasons.

"If I think the client will be slow to pay."

"For attorneys, I ask right away."

"A retainer is a must if the work will be on an infrequent basis."

"When consulting is on a continuing basis."

"Usually when work is not well defined."

"If it is a new client."

"Depends on the nature of the assignment."

"Only when a job stretches over a long period with many hours charged against it."

"When the customer wants my services for a year or more."

"I might consider a retainer if the job is of uncertain duration, magnitude, etc."

"Always if any testing is involved."

"When a contract will restrict or reduce my opportunities for other consulting jobs."

"Only ask an unknown plaintiff attorney for a retainer."

"Only if a collection problem is anticipated."

"Depends on type of service."

"If I have no previous experience with the client or the job appears to be large."

"Use only for 'on tap' services exclusively by telephone."

How much retainer do you ask for?

"In legal cases the amount depends upon the amount of money involved in the case."

"Strictly negotiable with the client."

"Usually based on size of job—$500 minimum, $2,000 maximum."

"An amount equal to five days of work."

"50 per cent of the anticipated fee."

"I use intuition to determine the amount to ask for."

"Estimate of costs and charges for one month."

"About 25 per cent of the monthly fee."

"May demand a higher retainer for giving legal depositions or for time in court—and may demand pay before I perform."

"Some clients set up a drawing account based on dollars per year."

"Local business, $500 to $1,000; out-of-town business, $1,000 to $1,500."

"20 per cent of estimated amount of fee up front."

"Equivalent to pay for a significant portion of the first phase of the work."

"Now considering asking for retainers when I work for lawyers."

"Currently $400 per month, arrived at through experience."

A man who does not use retainers "does some work on an open purchase order. I bill when work is requested. Some of my P.O.'s have a 'not to exceed' provision."

Another consultant points out, "Having a retainer tends to reduce the possibility of encountering conflicts of interest . . . because you are committed to a client."

Another comment, "Haven't asked for one yet. I was considering one on a proposal involving 60 days of work per year. I would have estimated expenses for one month and asked for that amount up front to cover such expenses."

The Minimum Job a Consultant Will Take On

Forty per cent have minimums. They run: " . . . half day . . . four hours at $250 . . . $200—this scares off some nuisance jobs . . . one hour is my minimum . . . $100 . . . one day . . . two hours . . . one day for a new client, less for a repeater . . . $50 . . . $500, and I throw in a final report."

About 60 per cent do not have minimums. They say: " . . . the deciding factor is my interest in the job. I like technical challenges . . . locally, no; if I need to travel by air, I charge at least eight hours or one day . . . I prefer short jobs over long ones . . . no, if I feel my services are necessary . . . if the job is too small, I prefer to give free service over the phone . . . may not charge if the job can be done in less than four hours and there are prospects for a larger job later on . . . the deciding question is whether I'll feel comfortable on the job . . . no, some of my billings are under $100 . . . small jobs may be a way of getting one's foot in the door, or to maintain warmth in relations with a client . . ."

Bidding, Estimating, Writing Proposals, Giving Presentations

Overview

To bid, you need to know something about estimating. Bidding is not a standard requirement, but a grasp of the basics is recommended.

A presentation is an oral version of a proposal with some rules related exclusively to public speaking.

One view on writing proposals:

"There are many books on the subject. Usually, I have found a standard format does not do the job; each job is unique, and the client looks for your input on his problem: what will be done, how long it will take, what to expect, how much it will cost, when you can start, how to be paid—all, of course, after the problem is defined and agreed upon and understood by all parties concerned. Don't promise specific results unless you can make a fix, do it, see it, and measure it. The result must be something tangible."

Tips on giving presentations:

"Be prepared; tell them what you are going to say, say it; tell them what you have said; keep it short and to the point; leave plenty of time for discussion; and be prepared to field any and all questions; but by all means do not try to bluff an answer to a question you really can't answer."

Some Do's and Don'ts When You Are Asked To Bid

Clients may ask for a bid—usually the government or where there is competition among consultants for the job. About 40 per cent have had experience with bidding. However, a typical reaction is: "Bidding is not regarded as professional."

You hear:

"I do not believe in bidding as such for consultants. If a client is looking for the low price, he does not want me. If he wants me, he trusts me."

"I have never bid. My rates are considered reasonable, and I do not make exceptions."

In case you are asked to bid: ". . . make sure all expenses are covered . . . try to be exact. But if you make an error, try to swallow it . . . bid in phases with estimated costs per phase . . . if you know there will be competitive proposals, you should find out if someone has it wired. If so, there may be no point in your presentation. Generally, I steer clear of the bidding concept . . . bid the amount that will return a profit for your work . . . size up any potential penalties and determine whether testing is involved . . . I always add about 40 per cent to my estimate and find it reasonably accurate . . . don't bid on anything unfamiliar to you . . . don't underbid just to get the job . . . there is a danger of not bidding high enough. Time adds up . . . "

Some miscellaneous comments:

"I can't give any tips on bidding. Each consultant must determine for himself how he proposes to approach prospective clients."

"Don't bid. Spend your time developing single-source clients."

Expert Advice on How To Estimate

The necessity to estimate on a job for your own information or a client's is common, while the requirement to prepare and present a proposal or to bid is not.

About two dozen ways of estimating are reported.

"Usually an estimate of time needed to do the job plus an hourly rate ($70) plus travel expenses."

"Past experience."

"Strictly on a per diem basis. I may guess total days."

"Take into account the complexity of the job, the amount of work required to solve the problem, or to meet the client's requirements."

"I work for lawyers, but I may be asked to estimate my maximum charge."

"Materials plus overhead plus actual engineering time and related costs."

"Estimate hours required, multiply by hourly rate, then I may discount, say 10 per cent, on a big job."

"Estimate hours."

"I outline requirements, estimate consulting time and outside charges, if any; and take possible contingencies into account."

"Forty years of experience."

"Experience and carefully."

"Most of my jobs are short. I quote daily rate plus expenses."

"Take into account the complexity of the project, its duration, dollars involved, and whether it is a cost reduction or capital program, etc."

"My judgment as to time needed for investigation and possible exemplar and laboratory costs, plus travel time—if out of the local area."

"Estimate number of days, use a daily rate worthwhile to me, estimate expenses, adjust total in terms of my degree of interest plus national or future impact of the work."

"Scope of the job, details required, along with degree of expertise required."

"My estimate is based on experience if the job is simple. If it is not, the number is negotiated with the client on the basis of the cost of doing specific things and with the understanding that the numbers are on the low side."

"Estimates are based on experience, generally in days under fixed conditions, allowing extras for further work."

"Time only."

"Time, materials, and expenses."

"On small jobs, estimating is a gut feeling (eight hours, usually). Sometimes I use a half-day minimum. Sometimes I charge an hourly rate to become acquainted with the problem, then make a bid."

"I quote representative lab charges and indicate hours of my time that might be needed . . . in a broad sense."

What You Ought To Know About Writing Proposals

Most engineering consultants are obliged, at one time or another, to write proposals—typically because a client wants what amounts to a summary of the consultant's overall plan up front, or there is a competition among consultants for a job and the proposal provides the prospective client a way of evaluating the comparative merit of each contestant's plan.

What, on the basis of experience, should one put into a proposal? How should it be written?

There is no single preferred way of getting the job done "properly."

"1) Set up an hourly rate. 2) Estimate hours required. 3) Multiply #1 by #2. 4) Add estimated travel expenses. Total of #3 + #4 = proposal."

"Use good organization and be creative . . ."

"Keep it short, to the point, and don't try to solve the problem in the proposal."

"Avoid generalities, propose a specific course or courses of action, outline exactly what you propose to do and why, and clearly document your price so the client clearly understands your plan."

"Write to the point and make the proposal understandable and brief."

"Be concise, quote a price commensurate with your knowledge and experience; don't compromise your talent to come up with a low bid."

". . . in clear, concise, good English."

"Should be concise and explicit about what you will do, but don't give away all your plans and solutions."

". . . a proposal must strive to answer the basic problem addressed. Do not use fancy or complex words to describe something when a simple word will do it."

"Should be concise but complete: a statement of the problem, method of attack, estimate of time required, estimate of cost."

". . . a good abstract is important."

". . . need a good summary at the end. Elements of cost should be itemized and clearly understandable."

"Be direct, simple, specific, brief, responsive to the request for a proposal or to the problem, logical (no non sequiturs). Avoid excessive claims, promises, convey sincerity and credibility. Graphics can help."

"Be at least reasonably honest."

"Be aware that you cannot predict where an investigation of the problem will lead you, and be cautious in estimating the time required. Make the client aware of the intangibles, although you must be as specific as possible without donating your technical know-how for free."

"Research the subject thoroughly."

"Simple, straightforward, easily understood language. Statement of the problem and plan of attack should be direct, but ways should be left open for changes—by mutual consent—as conditions change; keep it short and direct."

"Most proposals are simple one- or two-page letters with required attachments, e.g., resumé, fee expense schedule, definition of work assignment, and schedule (if not covered in letter), technical papers on the subject written by consultant, patents, etc."

". . . allow for expansions in the project, even going beyond the client's expectations. It may be necessary, for example, to double the amount of equipment presently under consideration. Propose the best of equipment."

"A proposal should include: 1) Your educational background pertinent to the problem. 2) Your experience related to the problem. 3) Expertise related to problem. 4) Estimate of time. 5) Fee and travel anticipated. 6) References, if appropriate."

"Break down proposal into points so the client can understand why the bottom line may appear to be high, and he can separate essentials from nonessentials."

Some extraneous comments:

"Writing anything technical is difficult, requires maximum in organization, clarity, etc. I do not think a person should be told how to write anything. Proposals fall into this category. The important thing is to keep the desired end in mind, and to make sure the reader understands where things are going."

"Avoid proposals, if possible. Never write a proposal unless you have positive assurance you will get the work. Otherwise, refuse to write anything. Refuse to be involved."

"It's important to know who will read and evaluate your proposal; make sure sufficient detail is given so that the proposal will be understood and appreciated."

"Don't write a proposal unless you are sure it will be considered equally with the others."

"You must have self-confidence to make firm recommendations; and you need to be a good salesman to sell your ideas after you have made your recommendations."

"What does the client really want? If you can't figure that out in advance (or find out), don't write the proposal."

"Study Strunk and White's *The Elements of Style*."

Saying It, Rather Than Writing It

In this context, the presentation is considered, in intent at least, an oral version of the proposal. Chronologically, it follows the proposal. Some but not all clients ask for them.

As in the case of proposals, consultant and client practices and preferences wander all over the ballpark. Formal. Informal. Long. Short. Many visuals. Few visuals. No visuals, etc. You can find support for any position you choose.

"Prepare a careful outline and then condense it into very brief notes for reference. Make sure to secure good visual aids if they are desired. Speak slowly and distinctly. Address your audience individually. Try to keep distracting arm and hand motions to a minimum."

"Know who the audience will be before you start to work up your presentation. This may take as much as a third of your preparation time."

"Use lots of visual aids—photos, slides, charts, Xerox copies; bring reference material and have pages tagged for easy reference when questions arise."

"Know your subject; if you do, you exude confidence. Prepare in advance: outline, use visual aids—slides, models, charts, artwork."

"You may want to pre-sell certain individuals before making your presentation; be prepared for those who may try to torpedo your ideas during the presentation; try to give others as much credit as possible."

"I usually talk off the cuff. Depends on the clients, but sometimes I make my presentation over the telephone."

"Be factual . . . know your stuff, and do your homework."

"Requirements are the same as those for a proposal."

"Be as professional as possible."

"Mine are all informal conversations or telephone contacts."

"Finish in the time allotted."

"Speak from knowledge, with notes if necessary, but never read a presentation. Use visual aids if they really illustrate the points, as with simple tables or figures. Sometimes the fuss of setting up visuals detracts unduly and should not be used unless they are really needed to get a point across."

"Finish with clear-cut recommendations. If recommendations are in the form of options with varying costs, many clients appreciate the opportunity to make the final decision—even though you may have led them into it."

"Make sure slides are adequate and the presentation understandable to those not expert in your discipline."

"Keep in mind it is more difficult to absorb information in listening than in reading if the subject is difficult. An occasional touch of levity helps. An impression of informal dialog rather than 'selling' or 'talking down to' is desirable. It is the personal touch."

"Work for a condensed summary on cards, with copious slides or photographs and diagrams. Make all tabular figure presentation sufficiently large and clear for reading legibility and understanding. Speak clearly and distinctly, according to the acoustics of the room."

"Physical models are frequently helpful, particularly in court."

"1) Be factual and businesslike, with a smattering of appropriate humor. 2) Hand out printed copies of slides prior to presentation so listeners can make notes on the copies and do not get distracted from what you are saying by feeling it necessary to take copious notes. 3) Encourage questions at any time."

"Make sure all principal parties are present . . ."

"Always communicate with the key individual involved; with a group aim for the average person. Use slides, drawings, etc., to be sure certain points you are making are comprehended. If the presentation is fairly lengthy (20 to 30 minutes), have an introduction, a general but detailed discussion, and then a simple summary. Do not be verbose. Time is important, to you and the client."

"Be extremely well prepared and have startling illustrations."

"Research subject well beforehand, dress neatly, have good slides and graphs, and present your material in a logical sequence. Be concise."

"I have enrolled in Dale Carnegie, Toastmasters, and other public-speaking seminars; also have taken technical writing and management training. Follow a simple and clear outline. Make brief and to-the-point presentations."

"Know your subject. If visuals are used, don't read them word for word. Most people can do their own reading. If there is a question you can't answer, say so. Don't try to bluff it out."

"Presentations need the same elements as written material, except more attention sustainers are required. Charts, slides, and outlines are appropriate, for example. Do not read a text unless you are very experienced at doing this. Reading is the best way to put people to sleep."

"Use simple language understandable to the layman. Don't try to 'snow' the audience."

"Don't use technical terms, don't talk down to the client, promise only the reasonable."

"Know the client's company, its products, and how its organization works. Know what you can do and how your expertise will help resolve the problem, etc . . ."

"Go very easy on hype. Be sure what you say is current and understandable. Be sure what you say is related to the problem. Make sure your presentation is adaptable to the conditions where it will be given."

Chapter 23

Ground Transportation, Paying Bills, Work Out of State or Country

Overview

When do you drive your personal car on business? Rent one? Lease? Use plastic money to pay expenses? Checks? Cash?

Any special considerations, including fee practices, for work out of the state? Out of the country?

Should you ever take your wife on trips?

How You Get Around at Home, Away From Home

The rule of thumb seems to be: If the job is within comfortable driving distance of home or office, use your own car. Otherwise, rent one. Some consultants have company cars. Leasing does not seem to be a common thing.

Some practice reported:

"Use own car in state; rent one out of state."

"Locally, my own; when I fly, I rent a car."

"Own car most of the time."

"Use own when possible; but may have to rent when I am out of town on business."

"Use own car within 300 miles of home. Rent when I fly to other cities or countries."

"Rent when I fly if the client does not pick me up at the airport."

"Use my own car and charge it to my business."

"Company car."

Credit Cards, Cash, Check, Etc.

"I use credit cards for record-keeping purposes," says a consultant. "It's an excellent way of keeping track of your expenses. Receipts for everything. Can't have enough."

About 65 per cent of his colleagues concur, while 26 per cent favor checks, and 14 per cent vote for cash.

The primary motive for a record is that consultants need "proof of expenditures if challenged, especially by the Internal Revenue Service."

Typical practice is to use a combination of plastic money, checks, and cash (in that order of priority). The first two for major expenditures, and the last for small purchases. Credit cards have special convenience in traveling.

What's Different About Working Out of the Country

About 40 per cent of the panel have had experience in working out of the country. They advise:

"For jobs in Canada, I ask to be paid in U.S. dollars."

"I hear that Canada does not present any special problems for the visiting consultant, while Latin America does."

"Who takes care of clients while you are on the trip?"

"Language, funds, and translation services are possible problems."

"I have restricted overseas travel to six weeks at most. I have turned down some interesting jobs because being away is too disrupting to my practice."

"I only present technical papers out of the country."

"You run into delays in getting reports typed, and have to watch translations closely."

"Remember, doing anything overseas seems to take more time than it does at home."

"Transfer of funds [your pay] to the United States can be slow, and problems may arise."

"You need someone in your office back home to answer the telephone, maintain necessary correspondence, and keep in touch with certain clients."

"I ask for advance payment in U.S. dollars. To get along in a foreign country you need to be flexible and capable of adapting to local customs and working conditions. Most consultants doing foreign consulting set up firm return dates in advance. Of course, you can always back out of a job if you want to, but it is hard to terminate an indefinite agreement."

"Learning and understanding the culture of the people you work with overseas can be a problem. Please read *The Ugly American* beforehand."

"I consult for a foreign government on a no-fee basis. They cover travel and living expenses. I volunteered for the assignment."

"One must be conversant with local customs and ways they are dealt with, currency values, how to communicate, import/export regulations, time differences, etc. The biggest problems concern logistics in relation to distance from home; it is often hard to communicate; and person-to-person interaction is expensive but necessary."

Some financial aspects of working out of state and out of the country deserve separate consideration. Most consultants charge more for out-of-state and foreign work than they do for the local variety. Time away from the office and travel time are major considerations.

No unusual financial problems are encountered in work out of the state or in Canada. By comparison, in foreign work you hear, for example:

"Out of state, collections can be a problem; out of the country, impossible. For such work we ask for prepayment."

"When you are out of the country, you are always faced with currency exchange rates. I use rates in effect at the time a given expense is incurred. I always bill in U.S. dollars and request payment in same."

"In work out of the country, the contract should stipulate reimbursement for all out-of-pocket expenses."

"No personal experience with foreign work, but I would want to work through an American company."

"Mode of payment and international rates of currency exchange should be taken into consideration."

Some fee practices for out-of-state and out-of-the-country work:

"I tried the lawyer trick of one-third more money out of town, but it works well only if the job covers an extended period—in my experience."

"My only experience was with the International Executive Service Corps in Malaysia and Colombia. I was paid expenses and per diem."

"I have local fees; out-of-state fees, which are higher; foreign fees, which are at least double local fees; and fees in uncivilized, revolutionary countries, which are four to five times the normal fee."

"Out of the country I fly first class and charge more than I normally do for travel time."

"In foreign work I obtain my complete fee and expenses in advance."

"I ask about provisions for my wife to come along when travel overseas is extended. If the answer is negative, I obtain extra compensation for the long absence."

"Make sure an allowance for foreign travel is adequate; if the job is a long one, increase your fee."

"Your invoice should clearly state, 'U.S. dollars.'"

"Ask for an advance and for insurance coverage."

"I charge essentially full-time for travel outside my home state."

"Ask for an expense account for work out of the state or out of the country."

Many Opportunities for Wife To Travel With You

About 60 per cent state something like, "I prefer to have my wife accompany me if it is not a situation where she will get bored."

On the negative side you hear comments like, "No, I would discourage it usually . . . the work is No. 1, and she would be left alone a great deal of the time; 12- to 16-hour days are frequent—working, traveling, entertaining."

Advocates of the practice comment:

"It's good for her, and clients often enjoy meeting her."

"It makes business a pleasure. In addition, she has helped me take photographs, carry supplies, etc."

"Occasionally, as to Australia. Of course, I pay her expenses and make sure client knows it. Having the wife along is acceptable if it isn't overdone."

"Take her along, but not really on the job. She is on the trip to mix and mingle after-hours and at dinner meetings. Keeps her informed. She does most of my typing of reports, etc."

"Very occasionally, as a photographer, but that kind of teaming leaves an impression of threadbare consulting, and isn't likely to sell."

"Almost never on the job, but I may to technical conferences and meetings."

"My wife has come on about half of my trips, especially the longer ones. And while it makes me a little less flexible in my dealings, it has been a good practice. I would not do this in making an initial visit to a client because requirements are not yet known."

"I have had her join me after finishing a job."

"Many of my jobs wind up in court, with me trying to convince a jury of my conclusions. My wife makes a perfect juror. She can and will tell me if she is convinced by my presentation—or what I need to do to improve it."

"She frequently goes on trips with me and sometimes to meals. I see nothing wrong with it. It depends very much on the job."

"I haven't done this yet, but see no problem if she would enjoy herself and have enough to do. She goes to some conventions with me."

"Whenever possible . . . makes the job more pleasant and improves family relations. This should be done at the expense of some loss in profit."

"I always take my wife along on trips. She is on her own during business hours; unless, for example, she gets together with the client's wife."

"She comes along whenever she can get free and the trip is interesting to her."

"She stays out of sight."

"I traveled so much alone for so many years that now I never leave home without her. She is very personable."

"A little now. She will frequently when my last child graduates from high school."

"She loves to see what I work on, where, and meet the people. I do this only on established jobs."

Chapter 24

Joint Deals, Client Credit Ratings, Do's and Don'ts

Overview

Joint deals with other consultants are "no problem if arrangements are defined up front."

If you have not had previous experience with a prospective client, you may want to investigate his credit rating—for example, in *Dun & Bradstreet* at your local library, or by conferring with another consultant who has worked for the company.

Among the do's and don'ts cited by established engineering consultants:

Do—"be well organized, reserved, have a sense of urgency and responsibility for the client's problems and finances—the same standards you observe for personal matters; and be a shirt-sleeve problem solver."

Don't—"oversell or overbook your time, play politics, or be used, or let resistance or resentment among client personnel stand in the way of data gathering."

Joint Deals Often Make Good Sense

"They are very important; they extend your expertise."

"No, I don't do it."

The former is the position of 75 per cent of the panel, while the minority says:

"I shy away from joint deals because I prefer to be on my own and assume responsibility for only my work."

"Have no desire to work a joint job, but I have recommended other consultants where the work is outside my field."

"I am not interested. However, if I feel someone is better qualified for the job than I am, I recommend him to the client."

"I avoid such arrangements."

"Clients can always call in another consultant if a second opinion or different viewpoint is wanted."

Those in favor of the arrangement say:

"When I work with other consultants, I have a specific understanding about charges and who pays whom. I do not believe in any kind of kickback."

"Have been involved in just a few of these deals. They worked out well."

"OK, but one person (me) must be the prime contact with the client and provide the final review."

"I welcome them, but they aren't strictly joint efforts for me. One consultant subcontracts for the other."

"No problems with such arrangements. As in the medical profession, there is room for both generalist and specialist. In many instances both may be required."

"Several associates with specialties work directly for me—on a cash basis, or exchange-of-information basis."

"I sometimes recommend a consultant in a related field whose expertise I need, but I have him make his own arrangements with the client for pay. Occasionally, I hire another consultant."

"Useful in certain cases, as where an unexpected complexity calls for special, additional expertise."

"It is fine if you know them and can trust them."

"No objections to the practice if your expertise is complementary, and he has a good reputation."

"Have worked with many, but one must watch out for sharp operators."

"Other consultants often have facilities I don't; I assist others, if asked, at minimum rates."

"Only if the other person is known to me. I often split a teaching job with others."

"The agreement must be clearly written up in a contract. Otherwise, you can lose a friend or money."

"I have subbed on many government proposals on a daily rate basis. If your name and bio are used in a proposal, get a clear agreement on the number of days you will be guaranteed on the job."

When and How To Check a Prospect's Credit Rating

There is no need to investigate, it is generally held, if the client is, for example, a "defense contractor, government agency, or major corporation, or an established law firm." The same applies "if the

114

client is well known to me, or the referral came from another consultant."

In case of doubt, you can always "look them up in *Dun & Brad-street,* or one of your clients may know their reputation for fiscal responsibility in the business community." If doubt still lingers, you can always "ask for money up front if you still want to take the assignment."

Other practices:

"Check other consultants and act according to their counsel. I feel this is a continuing problem."

"I inquire into the matter if the contract is worth more than $10,000 to $20,000."

"If the prospect is not known to me, I ask others who have worked for them."

"I go back to the person who referred me to the client."

"Better Business Bureaus will help you with this."

"I may ask a client I do not know for credit references—a bank and two other businesses."

"If it's an attorney, I ask attorneys I know."

Rules To Consult By

1. Have integrity.
2. Know senior management.
3. Maintain contact, but don't pester.
4. Listen and let the client participate.
5. Communicate, communicate, communicate.
6. Do, by all means, turn down any work for which you are not eminently qualified.
7. Don't be a know-it-all.
8. Don't jump to conclusions.
9. Don't sell yourself short.
10. Don't compromise your ethical principles for anyone.

These ten do's and don'ts and all other examples that follow radiate from a central theme: An engineering consultant is a "professional at all times."

A consultant should:

• Try to learn something about the lawyer who retains him. Does he pay his bills? Does he do his homework on technical legal cases? Does he make a concerted effort to understand fully what the consultant is going to state on the witness stand?

- Most clients have not properly defined their problem. Don't take their word as to what happened. Do your own evaluation of the problem. After all, if the problem is correctly defined, the solution is defined. This is your most important task.
- Do only what has been requested, and always allow time to complete a project.
- Establish a reputation in one or more specialties.
- Be known by your publications; cultivate business by maintaining confidentiality and refusing to discuss work for others with a client. Be sure of facts and confirm discussions in writing.
- Go to conferences and seminars—as an attendee and as a speaker.
- Make sure you properly prepare for a job; take a comprehensive approach to the problem; take other viewpoints into consideration.
- Keep a log of daily activities, item by item, in sufficient detail so they will be meaningful when you refer to them later on; follow up on leads in a timely manner; spend time to become fully acquainted with the problem; minimize unnecessary work.
- Have a clear understanding in advance about pay.
- In product liability cases, point out the desirability of making your work lead to a settlement. Be sure to get all statements and depositions others have made and study them carefully.
- Keep overhead low and run a tight ship.
- Be sure the job is complete and the client is satisfied.
- Try to provide that extra which shows you care about doing a good job for the client.
- Be honest. Get all the facts before you act.
- Keep in contact with your client—let him know costs and what you can do to benefit him.
- Do all you can to enhance your reputation; charge properly and consistently for your services.
- Try to guide the problem solving so it appears the client solved the problem.
- Show concern for and interest in solving the client's problem.
- Keep yourself educated in your field of expertise.
- Be willing to listen carefully; try to accumulate all the facts you can before you start to work on the problem; promise only the reasonable.

Don'ts often have their origins in diplomacy or ethics:

- Never belittle work already done by the client's employees or suggest they are wrong. Frame your proposals as alternative interpretations of data or experimental procedures.

- Don't bite off more than you can chew; don't commit yourself to so much work that the job can't be done properly without undue mental stress and turmoil.
- Do not enter into a 50-50 partnership.
- Don't just tell the client what you think will please him; give him the benefit of your best opinion. Don't be too positive. It is rare when you can be assured of being 100 per cent correct.
- Don't mislead a client about anything.
- Don't stand in the way of data gathering.
- Don't discuss details of one client's work with another client. You can give the other client's name as a satisfied customer.
- Don't push yourself too much at the beginning of a job.
- Don't be afraid to tell clients things you know they don't want to hear; don't make snap judgments.
- Don't underestimate your worth; don't sell your knowledge cheaply.
- Don't make the client feel like a dunce.
- Don't be pushy; don't insult the client or his staff.
- Don't be an aggressive wise guy; the people you work with will resist you.
- Don't use technical language with the client.

Chapter 25

Consultants: From the Client's Point of View

Overview

So far, the platform has belonged exclusively to consultants. Let's hear from another source: the people who hire them in industry—the clients.

The following questions reflect the scope of the topic:

Where do you get leads on consultants?

What do you look for in the first meeting?

Do you ask for resumés? What importance do you attach to them? Key items of interest?

Do you ask for written proposals? Key items in them?

Do you ask for written or oral presentations? The importance you attach to ability to communicate orally? To write?

What, in your opinion, are the two or three major qualifications of first-rate consultants?

Are credentials of a prospective consultant ever checked out?

What do you look for in particular as a prospective client?

When do you use a consultant?

How is his performance judged?

Do you ever ask for written contracts?

How do you judge the reasonableness of a consultant's fee?

Did you ever fire a consultant? Why?

What is the importance of a consultant's ability to deal with your personnel?

What, in your opinion, are essential personality traits for a consultant?

How important are the dress and appearance of a consultant?

Who pays expenses associated with the first visit?

Where Clients Get Leads on Prospective
Consultants and Other Matters

Prospective clients get leads on consultants by:

- Word-of-mouth recommendations—from his previous or current clients or from consultants familiar with his work
- Hearing him present a technical paper
- Reading a paper or article written by him
- Becoming acquainted with him at a meeting, seminar, or conference

Less common methods of obtaining leads include directories of consultants, the Yellow Pages and other forms of advertising, or advertising for them. One client searches abstracts and follows up when an author looks promising by going to sources like *Men of Science*.

In first meetings or interviews with consultants, prospective clients look for evidence of:

- Knowledge in the technical field of interest
- Record of accomplishment
- Ability to grasp the problem and what is wanted by the prospective client
- Ability to communicate orally and in writing
- A nonabrasive personality
- Common sense
- Integrity

A consultant is written off at this stage if it appears, for example, that he has been terminated and looks upon consulting as a stopgap to tide him over until a permanent position becomes available.

Resumés may or may not be requested. In any event, they are rarely the sole basis for the "hire" or "no hire" decision. By comparison, authorship or references are rated as more reliable evidence of ability. Of course, a resumé can provide a way to spot and eliminate the grossly unqualified—for such reasons as an educational background in an unrelated discipline or a lack of experience. Items of interest in resumés include a listing of previous clients, descriptions of types of problems tackled as an employee or consultant, and documentation of accomplishment. Believability, or lack thereof, is an inherent weakness of resumés. Everyone knows anyone can be made to look good on paper.

On about every other job the consultant is asked to prepare a written proposal that includes one or more of the following items:

- Maximum price
- Completion deadline or timetable
- Approach to project
- Results expected
- Description of what will be done
- Work procedures
- Outline or plan of project
- Review periods and progress reports

Both written and oral presentations may be necessary as a prelude to being hired. The two methods of communication supplement each other, with each typically serving a different purpose. Aside from spelling out such items as objectives, procedures, schedules, accountability, and costs, the written presentation also reflects ability to write, understanding of the problem, and ability to organize, for example.

The oral presentation is regarded as a way to clarify or expand upon the written presentation, to get an idea of the verbal skills of the consultant, and to gain insights into his personality—such as ability to get along with others.

"The consultant may be a brilliant technician, but unless he can communicate reasonably well in writing and orally, he probably will be graded a failure," a client observes. Of the two methods of communication, writing ranks No. 1 in importance. In the words of one client, "A clearly written, understandable document is what you will work from." There are times when an oral report alone will suffice. But top priority is given to "clear, concise written reports that are not wordy or overly descriptive and set forth precise conclusions."

A first-rate consultant, clients say, has a diversity of talents, including:

- A detailed knowledge of the field of interest from the technical and sometimes marketing points of view, along with a record of success
- Ability to communicate orally and in writing
- Ability to motivate people
- Ability to work within an organization
- Integrity (intellectual honesty)
- Ability to solve problems
- Ability to get results expeditiously

The first time around, a consultant's credentials will be verified unless he is known to the client or is recommended by someone the client respects. The checklist includes proof of educational background, previous clients, and evaluations of performance.

A shortage or lack of in-house know-how is a classic reason for hiring a consultant. Consultants are also engaged to:

- Verify an internal program
- Buy experience
- Create an atmosphere of change—"by hiring a recognized authority"
- Get help on short-term projects where employees may have the talent but would have trouble meeting deadlines
- Obtain outside opinions because established groups tend to think along the same lines after long association
- Confirm or refute existing opinions
- Get a second opinion
- Complement or supplement staff expertise

Some companies hire consultants to serve as expert witnesses.

Performance against objectives and the meeting of target dates are among the guidelines for judging performance. Clients also cite:

- "Value of his solution to my problem"
- "His creative input, even if it is unexpected or uncalled for"
- "Timeliness of results"
- "Value received for dollars spent"
- "His ability to grasp the significance of the problem, solve it, and describe results"
- "Results versus the written agreement"
- "Results and accuracy"

Hammering out a written contract satisfactory to all parties can take more time than the consulting job itself; but more often than not, putting it in writing "is just good business." You can avoid conflict or confusion later on—misunderstanding of the objectives, the timetable, and costs—if you put them in writing. In addition, "a written proposal can act as an outline of expectations."

More selectivity is practiced where performance or contingency contracts are involved. When they are used, results promised are usually measurable and spelled out in detail. In some instances, a contingency or performance clause may be required when the consultant does not have a track record.

The reasonableness of a consultant's fee is often judged on the basis of going rates or what other consultants have charged for a similar assignment. The discussion usually takes place up front—before hiring. Techniques used include:

- Compare the fee with what the cost would be if your own engineers handled the assignment.
- Take into account the experience of the consultant and the results anticipated.
- Judge in terms of your need for a consultant.
- Estimate what expected results are worth to you.
- Compare with what you would pay your own top technical people for doing the job.
- Agree on terms—such as $400 per day plus expenses.

Between startup and completion of a job, consultants have been known to be fired. When it became evident "he knew less about the problem than we did," or "a failure to perform was accompanied by an escalating bill," or "results were not in evidence," or "failed to stay on schedule and/or failed to produce acceptable results in the early phases of the program." On such occasions it may mean, "We were not diligent in our search for a consultant or in our interviewing."

The ability of a consultant to get along with client personnel is extremely important.

"If my personnel do not trust the consultant, the effectiveness of his effort will be low and probably invalid."

"A consultant has to deal directly with your people. If he can't get along, he gets poor information."

"He must get his information from employees and his main impact will be on them. He takes the burden off me if he can communicate with them directly."

"Dealing with employees is less important when the consultant is here for only a day or so. For longer periods, it is another matter."

It is also important for a consultant to be:

- Honest
- Enthusiastic
- Perceptive
- A no-nonsense, straightforward type
- "Able to get along with me and not disrupt my organization"
- Well organized
- Calm
- Communicative
- Firm in purpose

- "Relaxed in manner in a way that breeds confidence"
- "Low-key—an outsider can't be pushy"
- Believable
- Trustworthy
- "Easy to get comfortable with"
- "Should have the same desirable traits as any employee in management"

As expected, promptness and diligence get top ratings as work habits. Promptness includes arriving at meetings on time and getting reports ready on schedule. Promptness and diligence in combination are implicit in the remark: "He must make me feel he is trying to get the job done and is putting in a full day's work." Further, "consulting is not a job for the lazy." In fact, "the consultant should feel the job is fun." He should also "be able to readily adapt to the client's work environment."

Consultants themselves seem to place a higher value on dress and personal appearance than their clients do. Among employers of consultants, what is appropriate tends to depend upon the assignment. It is felt that the consultant should adapt to his environment.

There are exceptions:

"If he is too sloppy, I would probably develop a lack of trust."

"In early meetings, he must inspire confidence; appearance is an important part of this."

"Appearance is very important for an expert witness in court."

On the first visit, the consultant may or may not have to pay his own expenses.

Clients who pay "cover all reasonable expenses—transportation, meals, accommodations, etc." Or "travel expenses plus a per diem stipend."

Some pay only if "there is significant travel" as opposed to a local visit.

The alternative practice is not to pay expenses until after the consultant is hired. Amounts and arrangements may be spelled out in the contract. Usually, "all transportation, housing, and food are covered." Of course, other expenses associated with the assignment, such as testing or the purchase of materials, are assumed by the client.

Finally, some cost-sensitive clients take this position: "If the job persists over an extended period, it may be wise to try to hire the consultant as a permanent employee. It's less expensive, and you get the benefits of his continuing input of ideas."

Chapter 26

On Being an Expert Witness: Some Basics

Overview

"Don't write or use categorical words or phrases utilizing catchall words like 'always,' 'unsafe,' 'defective,' 'insure,' 'guarantee,' and the like," advises Robert C. Onan, Jr., a St. Paul attorney with extensive experience in product liability cases. "A general rule," he continues, "is to report facts in writing, store conclusions in cerebral central storage until such time when all relevant facts have been developed and their integrity scrutinized. Then conclusions will more likely stand out as inferences from reported facts."

From other sources, additional do's and don'ts are presented, along with pertinent definitions of topics ranging from strict liability in tort and negligence to malpractice and privity of contract; some tips on giving depositions; and some views on the qualifications and duties of expert winesses.

Tips on Working for Lawyers in Lawsuits or Arbitrations

Mr. Onan suggests the following "Do's" covering the consultant/client relationship, and preparing and giving testimony:

- Determine who is retaining you and develop a written agreement covering, for example, the nature of the effort, the basis of compensation, monthly billing, advance payments, and the right to withdraw if any bill remains unpaid after 20 days.
- Determine the goal(s) of your efforts and document this early; insist upon an initial up-front fee and expense advance with any new client, regardless of who retains you.
- At the outset, inquire as to the correct format for your monthly bills and to whom they are to be sent. Avoid billing only on completion of the project.
- Insist that the organization retaining you furnishes continuing

124

updates of information and reports that may have an impact on your role, the propriety of that role, or the expense involved.

- Confine preliminary report writing to data gathering, leaving to a later time the inferences, findings, and conclusions that may be sought by the person who hired you. Master the significant differences among the words "fact," "report," "inference," "conclusion," and "opinion."
- Determine, as early as possible, whether your efforts are being measured against performance criteria.
- Emphasize firsthand experience and observations rather than reports prepared by others.
- Study the part, system, or product involved before making any commitment on findings and opinions.
- Use picture-creating words and phrases.
- Make a model of the product or system you are attempting to analyze.
- Insist on an opportunity to review all depositions, interrogatories, and court papers (pleadings).
- Consult with the attorney before he takes any depositions related to your effort.
- Use carefully prepared visual aids in presenting your testimony—models, transparencies, photos, charts, etc.
- Learn the fundamental rights of a witness.

"Don'ts," in addition to the one given earlier, include:

- Don't write reports unless you are requested to do so.
- Don't take photographs without having good notes to support them—where they were taken, under what conditions. Put an identifying mark on your photographs so they can be later identified.
- Don't fail to ask for help in an area where you are not expert.
- Don't take a case if you have any doubts about the integrity of the hiring party. There are many ways of investigating the matter.
- Don't assume the lawyer who hired you will understand the technical matters involved in your testimony. Explain these things to him carefully, especially where understanding is required for key statements, opinions, findings, recommendations, etc.
- Don't take anything to the witness stand in a hearing or depositions room unless you are subpoenaed to do so, or asked to do so by the attorney who hired you.
- Don't assume a deposition is less important than live testimony.
- Don't get too dramatic or overzealous in making your presentation.

Finally, a capsulized statement of what makes an undesirable expert witness: "One who is not knowledgeable, not willing to devote the necessary time to prepare, one who is argumentative; one who cannot clearly articulate his or her views and will not make a favorable courtroom appearance; or one who is concurrently working on behalf of claimants in other similar cases."

Some Legal Terms Worth Knowing

Legal nomenclature can be as foreign to the engineer as engineering nomenclature is to the lawyer. However, both parties need to have a nodding acquaintance with a handful or two of basic terms. In this instance, the number is set arbitrarily at 15: breach of warranty, case law, common law, compensatory damages, deposition, express warranty, implied warranty, interrogatories, malpractice, negligence, privity of contract, product liability, punitive damages, strict liability in tort, and tort. Except for the term "malpractice," the definitions are from a paper, "Products Liability and the Service Manager," presented by Prof. Richard A. Moll of the University of Wisconsin Extension at an Academy for Metals and Materials seminar entitled "The Expert Witness in the Metalworking Industry," sponsored by the American Society for Metals (Nashville, February 1983). The definition of "malpractice" is from *Architects and Engineers, Their Professional Responsibilities,* a book by James Acret (Shepard's Inc. of Colorado Springs, 1977).

Breach of warranty. Failure to keep an agreement or promise that certain facts were true, particularly as they relate to the performance or service of products. Promises may be written, oral, or implied.

Case law. Law generated by decisions of courts, particularly higher level courts (as opposed to legislative law).

Common law. Law originated, developed, formulated, and administered in England. It is not law created by legislature but principles and rules of action relating to government and security of persons and property, which derive their authority solely from usages and customs, or from judgments and decrees of courts recognizing, affirming, and enforcing such usages and customs.

Compensatory damages. Damages that compensate the injured party for injury sustained or to replace the loss caused by the wrong or injury.

Deposition. A method of discovering information related to a case and before the trial, consisting of questioning of a witness under oath. A court reporter or other authorized person records the questions of the lawyers and the answers of the witnesses.

Express warranty. Promise, usually written or stated, that certain facts are true. For example, a promise from seller to buyer relating to the product or goods, and is part of the basis of the bargain or sale.

Implied warranty. The promise or contract is not written or stated orally. For example, it is implied that the product or goods are merchantable if the seller is a merchant with respect to products or goods of that kind.

Interrogatories. A method of discovering information relating to a case before the trial. Answers to written questions are provided—such as detailed questions about important aspects of the manufacture and sale of a product. Interrogatories are regarded as a good and inexpensive way of obtaining important facts about the case.

Malpractice. Dereliction from professional duty or a failure of professional skill or learning that results in injury, loss, or damage. For example, the responsibility of an architect (or engineer in construction business) is essentially the same as that of a lawyer or physician. The engineer implies he has skill and ability sufficient to enable him to perform professional services at least ordinarily or reasonably well.

Responsibility in this context has been defined (Coombs v Beede 89 Mo, 187, 36A 104, 1896) as that of a person who "pretends to possess some skill and ability, and some special employment, and offers his services to the public on account of fitness to act in the line of business for which he may be employed . . . he is expected to perform at least ordinarily and reasonably well . . . and without neglect . . ."

Negligence. Conduct which falls below a certain standard established by law for the protection of others against unreasonable risk of harm. It is one theory a plaintiff can use in a product liability case, based on the fault or substandard care or behavior of a seller.

Privity of contract. The relationship between buyer and seller or between two or more contracting parties.

In a landmark product liability case in 1916 (MacPherson v Buick Motor Co., New York Court of Appeals, March 14, 1916, 217 N.Y. 382, 111 N.E. 1050), Judge Cardoza concluded that if a product is negligently made, is defective, and causes injuries, the manufacturer may be sued even though there is no privity of contract between the buyer and manufacturer.

Product liability. Case law (and some state statute law) involving the liability of sellers of defective products. The term is used to describe a type of claim for personal injuries or property damages arising out of the use of a product.

"Generally, product liability is limited to sellers of products, but almost anyone connected with the chain of distribution or sale, including those providing services, can be held liable," explained Professor Moll. He continued, "Whether the product liability case is based on negligence, breach of warranty, or strict liability in tort, there are four common elements: 1) A defect in the product. 2) The defect must have been present when the product left control of the defendant. 3) There must be injury or damage. 4) There must be a casual relationship between the defect and the injury or damage. The same criteria hold for service-related liability cases, if one inserts the word 'service' for 'product.'"

Punitive damages. Damages awarded to the plaintiff over and above what will compensate him for his damages where the wrong was aggravated by circumstances, or by wanton and wicked conduct on the part of the defendant.

Strict liability in tort. The main requirement is that a defect existed in a product when it left the control of the seller and caused injury. In the strict sense, the term is misleading in that it implies that fault is not a requirement. In the landmark case, damages were allowed for a breach of implied warranty; there was no privity of contract (Henningsen v Bloomfield Motors Inc. et al., New Jersey Supreme Court, No. A-50, September Term, 1959. May 9, 1960. 32 N.J. 358, 161 A. 2d 69, 75 A.L.R. 2d 1.).

Tort. A civil wrong, other than breach of contract, for which courts will award damages.

More Background for Consultants Doing Legal Work

"Fitness for use" and "accident reconstruction" are among the key terms of consultants doing detective work for product liability cases, advised Ralph Daehn, staff consultant for Packer Engineering Associates, in a paper entitled "Use of Independent Laboratories to Investigate Product Failures" at the ASM seminar previously cited.

Fitness for use, it is explained, is the fundamental concept of quality. It is applied to all good and services, and is "judged by the user, not by the manufacturer or supplier of the goods or services."

The concept has three parts: proper design, suitable materials, and acceptable manufacturing processes.

In failure analysis, the aims are to determine if a given part failed; and if it did, to isolate the cause of the failure. Accident reconstruction is a procedure for doing this.

"Accident reconstruction," Mr. Daehn stated, "is a systematic investigation of all the pertinent facts and information available surrounding an accident or incident. This includes the post-investigative procedures as well as witness statements of the accident during its occurrence and any pre-accident information that would bring together the actual accident causes and sequence and results.

"Accident reconstruction is useful for many purposes. First of all, it may be necessary to determine the actual sequence of events or movements of various components which occurred during an accident so that the proper failure mode that causes the accident can be determined. Determining accident causes establishes the responsibility for the cause of the accident. This is then useful in reexamination of design, operating procedures, etc. that would be useful to prevent recurrence of the accident in the future by redesign, different materials, improved manufacturing processes, new operating procedures, better warnings, new safety devices, etc."

Giving Depositions, and Obligations of the Consultant/Witness

More detailed information on giving depositions was presented at the ASM seminar in Nashville.

A deposition is a legal proceeding conducted with some rules of court. The aim is to gather and preserve the testimony of a witness for use in court. The usual site is a lawyer's office. Those present include: the consultant (witness), a notary public (to administer an oath), a court reporter (often the notary) to record testimony, lawyers for all parties in the lawsuit, and, if they so choose, the parties themselves or their representatives.

After the consultant takes the standard oath for a witness, lawyers take turns asking him questions. The court reporter records all questions and answers. They are later typed and bound in a book called a transcript or deposition.

The proceedings are fairly informal, but the informality can be misleading. Depositions are vital to a lawsuit and the attorneys are serious.

The lawyer for the plaintiff uses the deposition as a way of discovering what the consultant knows about the case. He is looking for evidence favorable to his side; he tries to commit the consultant to statements, and he may be looking for ways to discredit the consultant's testimony. For example, he may try to get the consultant to make statements that conflict with testimony of other witnesses in the case.

The lawyer for the plaintiff also will attempt to determine how the consultant's side plans to defend itself in the lawsuit. In some instances, a lawyer may take a deposition as a way to preserve testimony for a trial—in case the witness becomes ill or otherwise unavailable; under these circumstances, the deposition may be read to the jury.

At all times, the consultant/expert witness should tell the truth; he should be fair, in the sense of being objective and not giving the appearance of favoring one side or the other; and he should be accurate. Additional tips:

- Never volunteer information.
- Make sure you understand the question. You may ask the attorney or court reporter to repeat the question as many times as necessary.
- Take time to think. Listen to the whole question; consider the question carefully; think through your answer; then state your answer concisely.
- Never guess. If you don't know the answer, say so.
- Don't be afraid to admit you can't remember.
- Be patient.
- Never lose your temper.
- Be polite and professional in answering. Don't be cute or sarcastic.
- Speak clearly.
- Always finish your answer. A skillful trial lawyer can sometimes cut a witness off in the middle of the latter's answer by quickly interjecting another question designed to lead the witness off in another direction.
- Correct prior answers if necessary. You have the right.
- Don't say you are familiar with a document unless you know it in some detail. Should the examining attorney refer to specific content in a document, delay answering until you are given a copy and have the opportunity to check it out for yourself.
- Be precise in answering, but don't be overly technical.
- Always ask in advance what documents you should bring.
- Be consistent. Some lawyers ask the same question many ways, hoping to trip up the witness.
- In giving a deposition, you have a right to be comfortable. The room may be too hot or too cold, for example. Say so.

So, You Are Going to Testify as an Expert!

This subhead was also the title of a presentation given by Dr. T.J. Dolan, professor of theoretical and applied mechanics, University of Illinois, at the ASM seminar. An abstracted version of the presentation follows, touching upon what the plaintiff must prove, the role of the expert witness, qualifications for an expert witness, some do's and don'ts, testifying, and cross-examination.

"The plaintiff's case," explains Professor Dolan, "usually centers around the claim that the product was in a defective condition—unreasonably dangerous—when it left the manufacturer's hands, and was the proximate cause of the accident. What is unreasonably dangerous or defective is often a matter for a jury to decide." In addition to other testimony, numerous charts, photographs, purchase records, calculations, and drawings may be presented in evidence.

The "expert" has special knowledge or skills "not ordinarily possessed by the average person. He is permitted to state his opinion and interpretation of the information to the court and jury. Some of these opinions may be based upon facts that he observed and testified to, or as is often the case, he may be asked to state an opinion based on a hypothetical question . . . the expert does not pass on the truth of the testimony in the hypothetical questions, nor the truth of the testimony of other witnesses, and his opinions are purely advisory. The jury may consider the expert's qualifications, weigh his reasons, and reject his opinions entirely if, in the judgment of the jury, the reasons given are not convincing or sound."

Speaking of qualifying or establishing the expert, Professor Dolan points out that some judges and some attorneys do not look beyond academic records, while in most cases experts earn their reputations as such after college. Evidence of accomplishment includes development of products, inventions, technical publications, leadership in technical society or standard-making bodies, special awards, and the presentation of papers and distinguished lectures.

As guidelines for testifying, nine do's and don'ts are offered:

1. Maintain a cheerful, calm, factual, and cooperative attitude.
2. Do not get angry or argumentative if a lawyer tries to browbeat you; retain your composure.
3. Be honest, truthful, and sure of the arguments made; do not exaggerate.
4. Be sure of the question that is asked or have it clarified if you do not understand.
5. Think before answering and answer the question directly.
6. Do not reach premature decisions—get all the available facts before committing to an opinion or conclusion.
7. Keep careful records, documents, pictures, parts, observations, calculations, etc., and remember you may refer to your notes about them in answering any question that may be posed to you.

8. Do not give voluntary information—answer only the question asked.

9. Say so if you do not know—you may even wish to refer to others who will be called for testimony as being in a better position to answer the questions.

Professor Dolan adds that the expert should school attorneys he is working with on the technical aspects of the case and the technology involved; if he feels the client does not have a case, the expert should advise a settlement.

Not all experts enjoy their turns on the witness stand. "Many qualified technical people," points out Professor Dolan, "find it a very frustrating experience to be confronted with the legal rituals of the courtroom, and their inability to put in layman's terms (for the benefit of the jury) their profound knowledge of the subject. On the other hand, they may find it even more disturbing to learn that the 'expert' for the opponents has convinced the jury that the other side's expert opinions are erroneous."

Speaking to clients, Professor Dolan advised, "It is good philosophy not to hire an outside expert who will testify for you . . . he must not be an advocate—not for you, nor against you. He may in some instances tell you and your attorney things about his findings or analysis that you aren't going to like to hear. But it is better to be realistic or be forewarned than to have the opposing counsel spring a surprise on you in the courtroom because you hired a 'yes' man rather than an impartial expert."

He adds, "Some experts tend to become advocates through the influence and prodding of a smart trial lawyer who is doing everything possible to win his case. By warping a few facts, or by avoiding the possibility of alternative interpretations, etc. an 'expert' can sometimes interpret the information in a manner favorable to his client."

In Professor Dolan's opinion, the greatest potential pitfall of an expert witness is the tendency of some to make an exaggerated statement or to attempt answers to questions outside his field of special competence.

Further, the expert must prepare himself for skillful cross-examination. "If you are going to be an expert in the courtroom," advises Professor Dolan, "you'd better prepare yourself by reviewing pertinent literature, codes, and specifications because the attorney on cross-examination may be sitting there with textbooks and other documents and ask searching questions. A favorite is: 'Do you regard so-and-so as an authority in his field?' Once you answer,

'Yes,' you are going to hear passage after passage from his book or articles that may be taken out of context. You may be made to look like you are disagreeing with the expert you recognize as an authority in the field. Be careful of recognizing anyone as an authority. The best and most truthful answer is, "I recognize certain things he has written as being authoritative, but I disagree with him on other things. This is the truth. I don't know of any engineers who agree on everything."

Chapter 27

On Being a Part-Time Teacher: Some Basics

Overview

Counting preparation time, portal to portal time for presenting a session, and subsequent time needed to grade papers, the engineer/consultant/teacher is not well paid in money.

Offsetting compensation is in the form of priceless job satisfaction, particularly that which comes in knowing one is fulfilling his responsibility to pass along knowledge and know-how unique to his profession.

In the consultant/teacher's favor is the fact that he has never really been out of school, because he is in a knowledge-oriented profession with a continuing need to keep current. He needs help at the practical level—in establishing himself as leader-teacher and in learning how to survive in the classroom environment.

Engineers Have Foundation To Teach in the Classroom

"Teaching experience" is one of eight items in the resumé for adjunct faculty used by the American Society for Metals. The importance of experience can't be discounted, but it should be pointed out that the engineer with no formal teaching experience is by no means a raw recruit. He may be lacking in classroom rituals and know-how, but even on his first assignment as a part-time teacher at the high school, night school, trade school, or university level, he usually won't have too many rough edges showing.

The engineer, by inclination and by training, is an explainer, a natural teacher.

He is a communicator. Writing is part of his job. Speaking is part of his job. He has had analogous experience presenting papers, making in-house presentations, etc.

He is familiar with the tools of the trade—the film and slide projector, the overhead projector, the blackboard, the flip chart.

He understands what it means to be a student because he is a student himself and has been all his career—reading and attending seminars and conferences in his never-ending effort to keep up with developments and trends in his field.

He is people oriented, openly and honestly enjoying association with individuals and with groups.

He is dedicated to his profession, which gives him that extra quality required to be a good teacher.

He has a deep sense of responsibility to his profession and feels a strong need to return some of what he has received from it.

Other qualifications of interest to schools are indicated by the additional items in the ASM adjunct faculty resumé: academic background, professional/industrial experience, research, awards, major publications, and society/association memberships (including offices held). In addition, the engineer is asked to specify "areas of teaching interest," is shown what's now being offered—a laundry list of 31 on-going courses—and is invited to suggest "other topics you would be interested in teaching in seminar form."

Typically, students, rather than other teachers or school officials, grade the engineer/teacher. For example, at the end of courses or seminars presented by ASM, students are asked to complete evaluation forms. Quality of instruction is the central theme of the first three questions: an assessment of the "entire program," "technical expertise of the instructor," and "effectiveness of the presentation." The program is rated on a scale of 1 to 16: 13 to 16 is excellent; 9 to 12, good; 5 to 8, satisfactory; 1 to 4, unsatisfactory. Scores may have a large influence on whether a course or seminar is repeated or whether an instructor is invited back.

Some Know-How for Getting About in the Classroom

Where the beginner is often lacking is in such practical techniques as how the teacher establishes himself with the class at the beginning, how to ask and answer questions, how to start and manage group discussions when the lecture starts to drag, how to cope with the dominant student who threatens to take over the class, and techniques for changing pace during the lecture.

The counsel that follows is from an excellent source, *Seminar Leader Guide,* which is distributed by two organizations—the Professional Education Center, 331 Madison Avenue, New York; and the University Seminar Center, 850 Boylston Street, Suite 415, Chestnut Hill, Massachusetts.

The first thing you have to do is to demonstrate to the class that you belong in the driver's seat, so to speak, because you know what you are talking about, you are organized and practical, and you like people.

"Establish your credentials in a quiet manner," advises the *Seminar Leader Guide*. For example, as members of the class introduce themselves, ask questions: "What is your area of responsibility?" "Do you have any topics you would like to have discussed?" "What do you hope to take away from this seminar?" Concentrate on learning names, company affiliations, and what companies do, etc.

Be yourself. Some teachers like to operate on a first-name basis. Others prefer formality. Either works, as long as it is evident that you are interested in and friendly toward the class. These things, along with being well prepared, will earn the respect needed to lead effectively and to create a proper atmosphere for learning.

"Questions are an integral part of an effective seminar." Make sure everyone hears the question. If there is any doubt, repeat it. If facial expressions or other signs suggest the question was too wordy or vague, don't hesitate to rephrase and repeat it. Police irrelevant questions. Be on guard for questions that misquote you.

"The response to questions is very important in the establishment of the student/instructor relationship." It's critical—to your authority as a teacher—to answer promptly, with little or no hesitation, and without hedging. If you don't know the answer, say so. Suggest sources or offer to look it up later, for example. Do not refuse to answer or evade. Do not guess—unless you say you are.

"Always be careful to avoid criticism and apologies." The idea is to avoid negatives. For example, don't start by confession: "I had a lot of trouble putting together my presentation. I hope you will learn as much as I did." Or "I must apologize for this dark slide." Or "I don't know why they didn't give us a larger room for this class."

"Instructors often have trouble generating group involvement and getting feedback." If you feel you are talking too much and not getting enough group interaction, stop. Ask questions—at random, if you don't choose an individual. This is a good place to stop and ask: "Does anyone have a problem understanding me?" "Anyone disagree?" "Have I overlooked anything of interest to you?"

Should silence ensue (15 seconds or more), ask an individual by name. Watch facial expressions, the half-raised hand, or glances at neighbors, or for some other indication of at least an embryonic response. Once the silence is broken, others will follow.

Listen carefully. The dialog will give you an idea of where you are getting through, scoring, and where you aren't. Guide the direction of the discussion, keep it on track, discourage lengthy and vague answers. Remember, don't be critical.

When group participation shows signs of winding down, it is a good technique to ask a series of short, quick questions that highlight or summarize the group discussion. This technique suggests another. As you come to a new topic on the agenda, preview it briefly—including, if necessary, its relationship to the preceding topic—then lecture, and end with a brief summary. In writing, this is known as "tell them what you are going to tell them, tell them, tell them what you have told them."

How effectively such techniques are practiced is another matter. For example, "Careful preparation of content and materials are key steps . . . but remember that knowledge of the subject is not knowledge of how to present the subject."

What is your platform manner? Are you rooted at the lectern, head buried in a script? Do you move about on occasion—from side to side up front, and up and down the aisles? Speak only from notes and maintain eye contact? Spend much of the time with your back turned to the class, lights low, facing the screen while torturing the class with endless grade C− slides and tables and hastily scribbled transparencies prepared the night before in your room at the Holiday Inn? Are you enthusiastic? Do you appear to be interested? Or in a hurry to get it over with and go home? Are you sincerely, honestly interested in the students? Do you care whether or not they learn anything? Appreciate the fact that ability to act and showmanship are important in teaching?

Tailoring the message to the audience is another key item in the teacher's bag of tricks. Usually, a balance of theory and practical know-how, preferably personal experience of the teacher, is required. In case it's a mixed technical/nontechnical audience, one practice is to concentrate on highlights, interpreting as much as possible for those who may not understand, and suggesting references and other sources of information for those who want greater technical detail; give them the opportunity, for example, to get together with you at coffee breaks and lunch. Handouts, such as copies of papers or articles, are also used as supplements. The overall objective is "solid presentation of the most current practices, recent developments, guidelines, tips, checklists, pitfalls—in short, practicality."

A tip: "Be familiar with new developments." Including, for example, significant publications, current practices, and contemporary trends in industry. "Most participants come to a seminar with specific ideas of what they expect to learn, and often are prepared with questions. Many pride themselves on keeping up-to-date."

Now to managing what takes place in the classroom.

For example, competing with the highly vocal, know-it-all student who tends to intimidate you and his classmates and eventually becomes a pain in the you-know-where. One trick is to make him work for, rather than against, you. Do this before the dominant individual becomes a problem. Turn the tables; for instance, call on him frequently to confirm or augment points you make. Ask him to relate his experience with the subject, his opinions. But keep control. Make it obvious to everyone that you are still the boss. Another trick, particularly if the dominant individual is a problem early on: Ask him to make a short presentation based on his experience and views at the end of the day. This will tend to keep him out of your hair the rest of the day as he thinks about and gets involved in preparation.

Another tip. Classroom seats are not upholstered. The length of the session and the uninterrupted drone of a lecture can trigger boredom, outbreaks of doodling, and blank stares into the wild blue nothing.

What's needed is a change of pace. "It's essential that you look directly at individuals in the class to attract and maintain their interest." This is something that can't be done if your head is buried in your script or you are addressing your remarks to a screen at the 1 o'clock position. Watch for signs: "When individuals start to look away from you, fidget, slump in their seats, or start to doze."

How do you shift gears?

"You should use various changes of pace about every 15 minutes to prevent loss of attention."

For example, start a group discussion, go from the lectern to a flip chart or to the blackboard, tell a war story, demonstrate something, give a quiz, distribute a handout, declare a bathroom break or a two-minute stretch break.

A couple of other techniques are in the acting/showmanship category: "Suddenly, dramatically, increase the volume of your voice; or vary the rate of your delivery; or walk about the room as you are talking." The last named item has another benefit. The class gets tired sitting in one position. If they follow your progress about the room, this will tend to ease the strain on their aching backsides.

Another tip: "Draw information out of the class. It stimulates them, adds valuable specifics, and saves your voice. If they don't come up with all the right responses, you can add them before going on to the next subject. Participants also like to be asked not only whether they are familiar with a technique, but what their experience has been. Use flip charts to record material you have drawn out of the class; the responses can then be referred to as is appropriate."

More how-to about group discussions. They can be planned ahead

138

of time. For your guidance, jot down representative topics and a logical order of discussion for later reference. Anticipate the possibility of little or no group participation at the beginning and come prepared to warm up the discussion yourself. Another technique is the cold approach. Suggest a good topic. Develop the outline via class participation, using the blackboard or a flip chart.

In running the discussion:

- Don't allow people to get into time-consuming and pointless arguments.
- Try to show the relationship of one topic to another.
- Try to get everyone to take part.
- Encourage the sharing of experience and ideas.
- Don't let one or two people take over.
- Make sure the discussion is headed where you want it to go.
- Flip the switch when the discussion starts to flicker.

A concluding key point: The teacher is responsible for staying on schedule. It's easy, for example, to let group discussions, questions/answers, unplanned detours, etc., sidetrack your agenda. Make sure you don't wind up without time to cover one or more major topics when you reach the closing minutes of the session.

Getting back to the overall view, the *Seminar Leader Guide* lists common mistakes made by beginning teachers. Among them:

- Appearing unprepared
- Improper handling of questions
- Apologizing for yourself or the sponsoring organization
- Being unfamiliar with the knowable—names of people in the class, their companies, latest issues of key journals, etc.
- Using audiovisuals unprofessionally
- Seeming to be off schedule
- Not involving all members of the class and apparent loss of control of the class
- Not establishing rapport and empathy
- Appearing disorganized; not previewing/presenting/reviewing each key section; not summarizing at coffee breaks, lunch breaks, or at the end of the day
- Not getting a strong start by establishing yourself quickly as the leader
- Being too theoretical

In student evaluations of engineer part-time teachers, common criticisms are:

- Lack of practical information—too much theory
- Material too elementary, out of date, or not state of the art
- Not enough group interaction—too much lecturing
- Failure to follow the brochure description of the course
- Sometimes boring
- Disorganized—skipping from topic to topic to topic with no apparent sense of direction
- Poor visuals

The cure in each instance, it is pointed out, is suggested by the nature of the complaint.

Chapter 28

Social Security, Record Keeping, Taxes, Legal Liability

Overview

The best advice is . . .

If you have a question about Social Security, call your local Social Security office.

If you have a question about taxes, call your local IRS office or visit a tax accountant or a lawyer with experience in this field.

If you have a question about keeping books, call your local IRS office or local Small Business Administration office or visit a tax accountant or CPA or a lawyer with experience in this field.

If you have a question about legal liability as a consulting engineer, visit a lawyer with experience in this field or call an insurance company offering such coverage.

For the do-it-yourselfer, reliable sources of self-help are available. Representative references are cited here, then highlighted in the sections which follow.

- For an overview of Social Security law: the *1984 Meidinger Guide to Social Security,* by Dale R. Detiefs, 12th Edition, 48 pages, published by Meidinger Inc., 2600 Meidinger Tower, Louisville Galleria, Louisville, KY 40402. Telephone: 502/561-4541. Topics range from payment of Social Security taxes by the self-employed to earnings limitations of persons receiving Social Security benefits.
- For an overview of the many roles of the small businessman: *Thinking About Going Into Business?,* Management Aids Number 2.025, 6 pages, published by the U.S. Small Business Administration (SBA). Available free by writing to SBA, P.O. Box 15434, Ft. Worth, TX 76119. Topics range from regulations and licenses to taxes.

- For an overview of services available through SBA at the state level: *The States and Small Business: Programs and Activities,* 237 pages, published by the SBA's Office of Advocacy, October 1983. Contact Office of the Chief Counsel for Advocacy, U.S. Small Business Administration, 1441 L Street NW, Washington, DC 20416. Telephone: 202/634-6098. Topics range from financial assistance to available publications.
- For an overview of record-keeping requirements: *Keeping Records in Small Business,* 6 pages, Management Aids Number 1.017. Available free by writing to SBA, P.O. Box 15434, Ft. Worth, TX 76119. Topics range from minimum records required to a simple filing system.
- For an overview of tax requirements: *Getting the Facts on Income Tax Reporting,* 4 pages, Management Aids Number 1.014. Available free by writing to SBA, P.O. Box 15434, Ft. Worth, TX 76119. Topics range from checkbook records to keeping records for fixed assets like office equipment.
- For an overview of legal responsibilities to the public and clients: *Consulting Engineering Practice Manual,* edited by Stanley Cohen, 195 pages, sponsored by the American Consulting Engineering Council, McGraw-Hill, New York. Topics range from legal relationships with clients to professional liability.

Each reference is reviewed briefly in sections which follow. But first, to round out this introductory section, brief answers to several common questions.

Q: The major benefit(s) of incorporating?
A: Generally, none now. Before a change in law, the main attraction was gaining the equivalent of a tax-free fringe benefit package—a retirement plan along with health, medical, and life insurance.
Q: Roughly what does it cost to incorporate?
A: At least $600—$500 or so for attorney fees and $50 to $100 in registration fees.
Q: Is incorporation a shelter for legal liability?
A: No more than it is for a medical doctor or lawyer.
Q: Is liability insurance recommended even if you incorporate?
A: Right.
Q: About what does insurance cost for an engineering consultant?
A: The American Society for Metals, for example, offers a plan for members who are employed engineers and consulting engineers. For the latter, the annual premium runs from $185 to

$578, depending on such factors as gross earnings and extent of coverage.

Q: Without being incorporated is it still possible to use certain business expenses as tax deductions?

A: As a self-employed person, yes.

Some Highlights of Social Security Law (Meidinger Guide)

What happens to your Social Security coverage when you choose to stop working for an employer and strike out on your own as a consultant?

When you retire at 65 and opt to take your Social Security benefits, how much can you earn without detriment to your Social Security entitlement?

The self-employed are covered by Social Security, reveals the *Meidinger Guide to Social Security.* Payment is based on net annual income if it is at least $400. Net income is figured under the same rules used for income tax purposes. The Social Security tax is computed on a special self-employment income tax schedule.

The total one may earn and still receive his or her full benefits is called the earnings limitation. New rules went into effect in 1984. Age is the governing factor. For those 65 or over, the amount is $6,960. For those under 65, the amount is $5,160. Both ceilings are subject to yearly escalation.

Until 1990, benefits are reduced $1 for every $2 exceeding the earnings limitation. Starting in 1990, the ratio will be $1 for every $3 above the earnings limitation.

Certain income is not counted in computing the earnings limitation, including:

• Pensions and retirement pay
• Payment from certain tax-exempt trust funds such as profit sharing plans, bond purchases, or annuities
• Dividends
• Rental income, unless one is in the real estate business

The self-employed are subject to the earnings limitation requirement until they reach age 70. Profits and losses of all businesses in which you are engaged are added together to determine earnings. If the total yearly profit is over the earnings limitation, your benefits will be reduced even if one or more of your businesses lost money.

Another change in law went into effect in 1984: Part of your Social Security benefits (but not more than 50 per cent) may be treated as taxable income. The penalty starts at $25,000 in other income for a single person and at $32,000 for married couples filing a joint return.

In addition, if you continue to work after retirement, you must pay the Social Security tax on the additional income.

There is an incentive for late retirement. By working beyond age 65, benefits increase incrementally each year through age 70. For those reaching 65 in the 1982-89 period, the yearly gain is 3 per cent. For those waiting until the year 2009, the reward is a yearly gain of 8 per cent.

The Consultant as a Small Businessman

"To start and run a small business you must know and be many things," reports the SBA in its brochure, *Thinking About Going Into Business?* "As one small business owner attending a conference put it: 'When I came here, my business lost the services of its chief executive, sales manager, controller, advertising department, personnel director, head bookkeeper, and janitor.'"

Representative topics discussed:

- Basic survival skills needed to run a business include a working knowledge of basic recordkeeping, financial management, personnel management, market analysis, break-even analysis, product or service knowledge, legal structures, and communication skills.
- Regulations and licenses vary by business and by state. Your best bet is to go to your local SBA office. Local chambers of commerce can usually help, too. For example, a business license from the city, town, or county may be required. In most states, if your business is not incorporated you will have to register under what is called the fictitious name law. You may also be required to file for a sales and use tax number. At the federal level, contact your local IRS office.
- Under the subject of casual labor and taxes, if you employ an individual, you must withhold taxes. If the labor is on an independent contract basis, the independent contractor withholds taxes and files appropriate forms. Say you have a partner and your wife or his wife keeps the books. Is she an employee? She is only if she is paid. If so, taxes, including FICA, must be withheld for her.

Help Available From the SBA at the State Level

The information in SBA's *The States and Small Business: Programs and Activities* is presented state by state. Programs for California, Colorado, and Massachusetts are summarized here to indicate the nature and scope of what's available.

California has two major offices, both headquartered in Sacramento, to assist small businesses. The Office of Small Business Development provides information on procedures, regulations, and licensing and serves as an ombudsman for small businesses. The Office of Small and Minority Business helps small businesses compete for state procurement and construction contracts and expedites payment by state agencies to small businesses supplying goods and services.

California's Department of Economic & Business Development includes three offices that offer services or assistance to California firms either directly or through local government units. Small Business Resource Centers provide counseling to small businesses that can't afford professional services in such areas as accounting, marketing, and economic analysis. The state also has a variety of loan guarantees and direct loan programs.

Colorado has an Office of Regulatory Reform in Denver "to provide one-stop business permit and licensing information . . . and to recommend the elimination of unnecessary, burdensome, and/or duplicating regulations." The office's Business Information Center provides information on federal, state, and local licensing requirements for people starting a small business.

The state has a Small Business Assistance Center based at the University of Colorado, Boulder, with branch offices in Grand Junction and Pueblo. They provide management assistance and counseling. The Office of Regulatory Reform also has publications. For example, the *Colorado Business Startup Kit* includes all federal and state forms needed to become an employer in the state. *Doing Business in Colorado* outlines regulatory requirements.

In Massachusetts, the Small Business Assistance Division of the state's Department of Commerce (Boston) was established to "provide a one-stop comprehensive assistance office for the small business community. They provide technical and management assistance, information on permits and licenses needed to start and operate a business, and information on sources of financing."

Small Business Development Centers, affiliated with colleges and universities, are located in Amherst, Chestnut Hill, Worcester, Salem, Fall River, Lowell, and Springfield. Their mission is to give management and technical assistance to new and existing small businesses. The center at the University of Massachusetts, Amherst, helps businesses "seeking nonconventional sources of financing." The Massachusetts Business Development Corporation, described as a state chartered, privately funded secondary, fixed-asset lender, is located in Boston. Services include working capital loans, leverage buy-outs, and real estate loans.

Records You Will Need To Keep To Run a Consulting Practice

"A good recordkeeping system," advises SBA in its *Keeping Records in Small Business,* "must be simple to use, easy to understand, reliable, accurate, consistent, and designed to provide information on a timely basis."

At the minimum, four basic records are required: sales records, cash records, cash disbursements, and accounts receivable.

"A petty cash fund should be set up to be used for payment of small amounts not covered by invoices. A check should be drawn for say $25. The check is cashed and the fund placed in a box or drawer. When payments are made for such items as postage, freight, and bus fares, the items are listed on a printed form or even on a blank sheet. When the fund is nearly exhausted, the items are summarized and a check is drawn to cover the exact amount expended. The check is cashed and the fund replenished. At all times cash in the drawer plus listed expenditures will equal the amount of the fund."

Keep a careful list of permanent equipment used in the business—items useful for a year or longer and of appreciable value. Show date purchased, name of supplier, description of item, check number by which paid, and amount.

A charge to expenses should be made to cover depreciation of fixed assets, other than land. By definition: items normally in use for one year or longer, such as buildings, automobiles, tools, equipment, furniture, and fixtures. Straight line depreciation is typically used. An automobile with an estimated life of four years, for example, is depreciated at a rate of 25 per cent per year.

Several copyrighted systems providing simplified records are available from most office supply stores. Many are listed in SBA's *Recordkeeping Systems,* SBB 15. A free copy may be obtained by writing to SBA, P.O. Box 15434, Ft. Worth, TX 76119.

A simple filing system is recommended: a large manila envelope for each month's paid invoices, tax returns, etc. Paid returned checks are stapled to invoices or tax returns. "The cash disbursement record, which shows date of payment and check numbers, makes it easy to locate paid invoices. It is good policy to keep records up-to-date. The best way to do it is by seeing that your recordkeeping is done daily."

Guidelines for financial management outline some key things that should be done on a daily, weekly, and monthly basis. Daily, for example, it is suggested that the small businessman should know his cash on hand and his bank balance, maintain a daily summary of sales and cash receipts, maintain a record of all monies paid out by cash or check, and make sure all errors in recording collections and accounts are corrected.

Records Required for Income Tax Purposes

"When the facts are in proper order, federal income tax reporting is greatly simplified," advises the SBA's *Getting the Facts for Income Tax Reporting.* "Proper recordkeeping procedures are the key to accumulating the information necessary for computing the return schedules and the tax."

The IRS prescribes no specific accounting records, documents, or systems. It does require maintenance of a permanent set of books of account or records that can identify income, expenses, and deductions. Records must be available for inspection by IRS officers.

If your business is a sole proprietorship, you can use a simple set of records to capture facts for income tax reporting. "Such a system consists of a checkbook, a cash receipts journal, a cash disbursement journal, and a petty cash fund."

A separate checking account exclusively for business is recommended. Personal business should be handled through a separate checking account. All receipts (income from consulting) should be entered in a receipts journal. Money collected for sales tax, for example, is not income. All funds paid out should be recorded in a cash disbursements, purchases, and expense journal. Best practice is to record each check you write on a daily basis. Voucher slips should be kept to document each expenditure from petty cash.

Even with the simplest bookkeeping system, a fixed-asset record is needed.

"If your business is a small corporation or partnership, your records must cover situations that do not exist in a sole proprietorship. In a corporation, records must show salaries paid to its officers and dividends paid to stockholders. The owner-manager, as an officer of the corporation, is responsible for filing an income tax return for the company and a personal return to pay income tax on salary and dividends received from the corporation. If the business is a partnership, an information return on Form 1065 is filed, indicating income or loss assignable to each partner. Each partner then files a personal return and includes his or her share of partnership income with his or her other taxable income."

Keep records used in filing your firm's income tax return. There is no fixed answer to how long records should be retained. "Ordinarily, the statute of limitations for such records expires three years after the return is due to be filed."

SBA closes with this advice on paying taxes: "The government wants you to pay only your legal obligation—no more and no less. This fact has been best expressed by the late Judge Learned Hand, who said, 'Over and over again courts have said that there is nothing sinister in arranging one's affairs to keep taxes as low as possible. Everybody does so, rich or poor; and all do right, for nobody owes any public duty to pay more than the law demands.'"

Finally, SBA says, "Unless your background includes bookkeeping or accounting, you should use outside help in setting up your records. An accountant can help you determine what records to keep and what techniques insure that you don't pay unnecessary tax."

Relationship and Responsibilities to Client, Professional Liability

"The consulting engineer's relationship with the client should be one of trust," states *Consulting Engineering Practice Manual.* The consultant "should be viewed as an advisor in whom the client has confidence."

The reference sums up the consultant's responsibilities to the client by stating that he is the "faithful agent of the client on engineering matters."

Responsibilities are on two levels:

1. To "protect the safety, health, and welfare of the public in the performance of his professional duties."
2. To observe rules of practice established by the American Consulting Engineers Council (ACEC), such as taking assignments only when one is qualified, disclosure of all conflicts of interest to the client, and maintaining the confidentiality of information—"shall not reveal facts, data, or information obtained in a professional capacity without prior consent of client, except as required by law or ACEC guidelines."

As to liability, "a consulting engineer is professionally and legally obligated to provide services reflecting competence in the engineering principles involved and diligence in their application . . . courts have ruled that consulting engineers have a professional liability for errors or omissions which harm the client or the public. Careful supervision of the work of subordinates and checking of the final work product are expected of all consulting engineers. Perfection cannot be guaranteed, but all reasonable care must be taken to protect the public and the client."

Potential liability can be great. "To protect the public and the financial survival of their practices, most consulting engineers carry professional liability insurance coverage," advises ACEC.

An Engineers Professional Liability Plan is available to members of the American Society for Metals, for example. Two of the programs protect consulting engineers. Premiums are determined by level of service provided (called final work product), gross annual billings, and insurance coverage. There are two levels of coverage: Option A, with coverage of $1 million per year; and Option B, $200,000 per claim with a limit of $600,000 per year of coverage.

Final work product falls into two categories: Category V, "report, opinion, technical paper, research project, or computer software"; and Category VI, "a design or specification which will be used for a manufactured product or part thereof, a custom fabricated product or piece of equipment or part thereof, or an industrial or manufacturing process."

Consultants "may not be eligible for the coverage if their final work product does not fall into either category," advises some of the small print. Likewise if you have "professional or technical employees," etc.

Say you are in Category V and gross annual billings are in the $70,001 to $125,000 bracket. The yearly premium for Option A is $450; for Option B, $265.

Say you are in Category VI and gross annual billings are in the $70,001 to $125,000 bracket. The yearly premium for Option A is $578; for Option B, $340.

For information on membership in the American Society for Metals and eligibility in the liability insurance program, write to Chapter and Member Relations Department, American Society for Metals, Metals Park, OH 44073, or call 216/338-5151. The insurance plan is underwritten by Crum & Forster Organizations and administered through Smith-Sternau Organization Inc., 1707 L Street, Suite 700, Washington, DC 20036.

For a source on ethics, see *Ethics in Engineering,* by Mike W. Martin and Roland Schinzinger (McGraw-Hill, New York, 1983). For example, Chapter 5 covers professional responsibility and employer authority. Topics include conflicts of interest and confidentiality. Rights of engineers are covered in Chapter 6.

Chapter 29

The Panel's Bibliography

Other than technical books, handbooks, codes and the like, what do consultants read and recommend for a consultant's library?

Members of the panel for this book were asked the question. Results follow. No attempt has been made to round out the collection (other than to include most of the references cited by the author in Chapter 29), because its particular value is in the fact that these are the recommendations of the panel.

- Bermont, Hubert, *How to Become a Successful Consultant in Your Own Field,* Bermont Books, 815 15th Street NW, Washington, DC 20005, 1978.
- *Canadian Consulting Engineer,* magazine, Southern Communications Ltd., 1450 Don Mills Road, Don Mills, Ontario M3B 2X7, Canada.
- *Consulting Engineering Practice Manual,* Stanley Cohen, editor, American Consulting Engineering Council, McGraw-Hill, New York.
- Deney, Wilson, *Effective Use of Business Consultants,* Financial Executives Research Foundation Inc., 50 W. 44th Street, New York, 10036.
- *The Expert and the Law,* periodical publication, National Forensic Center, P.O. Box 3161, Princeton, NJ 08540.
- *1982 Guide to Fees,* National Forensic Center, 6 Ashburn Place, Fair Lawn, NJ 07410.
- Howell, E.B., *et al., Untangling the Web of Professional Liability,* 3rd edition, Design Professionals Insurance Co., San Francisco, 1980.
- Hoyt, Sam, *Men of Metals,* American Society for Metals, 1979.
- *IRS Guide for Small Business,* publication 334.
- Martin, Mike W., *et al., Ethics in Engineering,* McGraw-Hill, New York, 1983.
- *Meidinger Guide to Social Security,* 12th edition, Meidinger Inc., Louisville, KY 40402, 1984.
- Philo, H.M., *Lawyer's Desk Reference,* Vol. I and II, Lawyers Co-Op Publishing Co., Aqueduct Bldg., Rochester, NY 14694, 1979.

- *Reader's Digest*, article, "You and Your Rights," 1980.
- *Reader's Digest*, article, "You and the Law," 1973.
- *SAE Directory of Automotive Consultants*, Society of Automotive Engineers, 400 Commonwealth Drive, Warrendale, PA 15096.
- *Getting the Facts on Income Tax Reporting*, Management Aids Number 1.014, Small Business Administration, P.O. Box 15434, Ft. Worth, TX 76119.
- *Keeping Records in Small Business*, Management Aids Number 1.017, Small Business Administration, P.O. Box 15434, Ft. Worth, TX 76119.
- *The States and Small Business: Programs and Activities*, Chief Counsel for Advocacy, U.S. Small Business Administration, 1441 L Street NW, Washington, DC 20416.
- *Thinking About Going Into Business?*, Management Aids Number 2.025, Small Business Administration, P.O. Box 15434, Ft. Worth, TX 76119.
- *A Selected Annotated Bibliography of Professional Ethics*, Center for the Study of Ethics in the Professions, Illinois Institute of Technology, Chicago, 1980.
- Stanley, C.M., *The Consulting Engineer*, John Wiley, New York, 1962.
- *1983 Tax Guide for Engineers*, Academic Information Services Inc., P.O. Box 23279, Washington, DC 20021.
- Tomczak, Steven P., Tomczak & Associates, P.O. Box 480530, Los Angeles, CA 90048:
  Making the Proper Decision: Sole Partnership, Partnership, Corporation, Consulting Report No. 1
  How to Get Prospects, Consulting Report No. 2
  How to Turn Prospects Into Clients, Consulting Report No. 3
  Negotiations and Fee Determination, Consulting Report No. 4
  Tax Savings for Consultants, Consulting Report No. 5
  Twenty Startup Projects to Do in Your Spare Time, Consulting Report No. 6

Chapter 30

A Directory of Engineering Consultants

One panelist declined the invitation to be listed in this directory, explaining "I want to start tapering off on the number of jobs I take per year." Another panelist lives in and works out of Ontario, Canada. But even when you subtract these two from the total, the other members of the panel listed, according to good authority, still probably represent about one-third of all metallurgical/materials engineering consultants in the United States.

A compact format for listings was selected for the directory because of space limitations. In a very general way, the write-ups indicate the scope of the talent on the panel. By no means, however, do they serve as a substitute for credentials.

- **Armantrout, Clo E.,** 445 N.W. 12th Street, Corvallis, OR 97330; 503/752-1893.
  Specialties: Uranium alloying and processing, processing rare and special metals, handling zirconium-titanium-hafnium alloys and refuse.
- **Austin, William W. (Dr.),** 3221 Birnam Wood Road, Raleigh, NC 27607; 919/787-6946.
  Specialties: Failure analysis and accident reconstruction, product liability claims, metallic corrosion and corrosion prevention, materials factors in design.
- **Badger, F. Sidney,** 5200 Armida Drive, Woodland Hills, CA 91364; 213/347-1078.
  Specialties: Metallurgical and mechanical engineering, aluminum ladders, chipped hammers, automotive failures, industrial accidents, product liability cases, material usage in chemical and biochemical instruments, corrosion problems.
- **Blickwede, D.J. (Dr.),** Suite 310, 437 Main Street, Bethlehem, PA 18018; 215/694-3051 (business), 215/681-1444 (home).
  Specialties: Research management, process and product applications.
- **Bolz, Roger W.,** Roger W. Bolz & Associates, 205 Wyntfield Drive,

Lewisville, NC 27023; 919/945-5695.

Specialties: Automation of manufacturing, material handling, product design for economic production.

- **Bratkovich, Nick F.,** 5015 Knoll Crest Court, Indianapolis, IN 46208; 317/291-6803.

Specialties: Welding engineering, metal joining processes and metallurgy, design and service evaluations, problem solving, destructive and nondestructive testing, quality assurance and failure analysis. Experience with commercial products, pressure vessels, aerospace and nuclear weldments and related codes, standards, and specifications.

- **Brick, Robert M. (Dr.),** 32552 Sea Island Drive, South Laguna, CA 92677; 714/493-9462.

Specialties: Generalist on structure and related properties as determined by processing, including each and every step preceding the final structure.

- **Burke, Joseph E.,** 33 Forest Road, Burnt Hills, NY 12027; 518/399-8423.

Specialties: Ceramic materials and processing, material selection.

- **Campbell, Hallock C. (Dr.),** 746 Fiddlewood Road, Vero Beach, FL 32963; 305/231-1732.

Specialties: Welding education, welding metallurgy, welding electrodes and fluxes for stainless steels, nickel alloys, dissimilar metals.

- **Craft, C. Howard,** 16156 Amber Valley Drive, Whittier, CA 90604; 213/943-2133 (business), 213/943-1314 (home).

Specialties: Metallurgical failure investigation, fracture evaluation, materials and process engineering, nondestructive testing applications, photographic documentation.

- **Devis, Joe,** Devis Enterprise Inc., 15 Brandon Place, Rocky River, OH 44116; 216/356-2668.

Specialties: Heat treating, material selection, steelmaking and processing, failure analysis, metallography, mechanical testing, brazing, welding of steel.

- **Dove, Allen B.,** 933 Glenwood Avenue, Burlington, Ontario L7T 2K1, Canada; 416/637-6254.

Specialties: Ferrous rod and wire, production removal of oxides of iron, protective coatings, cold work, failure, fasteners, plant layout and practices, finishing, concrete reinforcing materials.

- **Dyrkacs, W. William,** Lochaven Road, Route 3, Waxhaw, NC 28173; 704/847-6557.

Specialties: Physical/process metallurgy and product/process development of superalloys, refractory metals, and specialty steels; manufacturing technology, process engineering, and facilities planning; vacuum processes and systems; materials selection and

heat treatment; acquisition evaluation; failure analysis; expert witness.
- **Faas, Bernard P. (Barney),** 22907 Felbar Avenue, Torrance, CA 90505; 213/530-5664 (business), 213/325-8288 (home).
Specialties: Developing welding procedures and qualifications, quality control in field construction and fabrication shops, failure analysis, materials selection, radiographic interpretation. Primarily involved in heavy industry, such as refineries, chemical plants, power plants.
- **Fellows, John A. (Dr.),** 650 S. Grand Avenue, Suite 1200, Los Angeles, CA 90017; 213/626-8761 (business), 213/957-0049 (home).
Specialties: Metallurgical and science consultant to legal counsel in cases of product liability with particular reference to failure analysis and fractographic instances of fatigue, mechanical overload, stress corrosion, embrittlement, defective design, defective manufacturing, etc.
- **Ford, Robert G.,** Ford Consultants Inc., 3100 La Playa Court, Lafayette, CA 94549; 415/939-1910 (business), 415/939-4377 (home).
Specialties: Metallurgical and manufacturing engineering related to metal fabrication and finishing, failure analysis, design assistance, and trouble shooting.
- **Frey, Muir L.,** (summer) 5951 Orchard Bend Road, Birmingham, MI 48010, 313/646-5068; (winter) P.O. Box 211, Leesburg, FL 32748, 904/787-8876.
Specialties: Metallurgical engineering for automotive and farm equipment industries. Material selection and specification, metallurgical equipment and process selection and operation. Specialist in heat treatment of gears.
- **Fritzke, Gerald P.,** Metallurgical Associates, P.O. Box 5486, Walnut Creek, CA 94596; 415/934-1161 (business), 415/935-7086 (home).
Specialties: Failure analysis, corrosion, brazing and welding, heat treatment, wear resisting materials and coatings, surface finishing, electronic industry problems, electroplating.
- **Gonser, Bruce W.,** 1301 Arlington Avenue, Columbus, OH 43212; 614/488-2183.
Specialties: Nonferrous metallurgy and the uncommon metals.
- **Hanzel, Richard W.,** Hanzel Associates, 75 S. 6th Avenue & 307, LaGrange, IL 60525; 312/354-5890.
Specialties: International management services specializing in product development, design, approvals, manufacturing, overseas sourcing and manufacture, general operation and management of consumer product manufacturing.
- **Hartbower, Carl E.,** 5033 Cocoa Palm Way, Fair Oaks, CA 95628; 916/967-9425.

Specialties: Welding engineering, welding metallurgy, welding research; product liability suits involving brittle fracture and weld failure, mechanical metallurgy, fracture testing, failure analysis.

- **Hayes, Earl T.,** 517 Gilmoure Drive, Silver Spring, MD 20901; 301/593-0920.
Specialties: Titanium, zirconium, hafnium, and other nonferrous metals, mineral resources, metal fines.

- **Howe, John P.,** 5725 Waverly Avenue, La Jolla, CA 92037; 619/454-6582.
Specialties: Materials science and nuclear engineering.

- **Hurlich, Abraham,** 1465 Robin Lane, El Cajon, CA 92020; 619/448-6168.
Specialties: Selection and application of metals and processes, metallurgy and mechanical properties of steels, titanium and aluminum alloys, failure analysis, high-strength aircraft and ordnance metallic alloys, brittle fracture, corrosion, behavior and properties of metals at cryogenic temperatures, metallic-non-metallic-composite armor materials, mechanism of armor penetration, terminal ballistics of armor and armor defeating ammunition.

- **Kattus, J.R.,** Associated Metallurgical Consultants Inc., 810 Fifth Avenue N., Birmingham, AL 35203; 205/328-3004 (business), 205/879-1446 (home).
Specialties: Physical metallurgy, failure analysis, high temperature testing.

- **Kern, Roy F.,** Kern Engineering Co., 818 E. Euclid Avenue, Peoria, IL 61614; 309/688-5532.
Specialties: Steel selection, purchasing of steel, metallurgical processing of steel, failure analysis, design of construction and material handling equipment, design of production tooling, cost reduction, expert witness.

- **Kinzel, Augustus B.,** 1738 Castellana Road, La Jolla, CA 92037; 619/454-6037.
Specialties: Generalist/specialist in materials (metals, refractories, plastics) and their engineering use. Fields include atomics, chips, computers, pressure vessels, aircraft design, instrumentation (particularly medical) and research management.

- **Koebel, Norbert K.,** 1100 N. 7th Avenue, Maywood, IL 60153; 312/344-7796 (business), 312/344-8155 (home).
Specialties: Heat treatment of metals, furnace atmosphere control and instrumentation, furnace and generator design.

- **Lement, Bernard S. (Dr.),** 24 Graymore Road, Waltham, MA 02154; 617/894-5702.

Specialties: Materials engineering services, including materials selection, processing, and product performance; failure analysis and accident reconstruction; expert witness.

- **Luce, Walter A.,** 663 Banbury Road, Dayton, OH 45459; 513/434-8148.
  Specialties: Selection of materials for handling chemicals (corrosion); corrosion failure analysis; metallurgy, production, and use of high alloys and specialty castings.

- **McKnight, Larry E.,** METTEK Laboratories, 1805 E. Carnegie Avenue, Santa Ana, CA 92705; 714/549-0343 (business), 714/528-4780 (home).
  Specialties: Materials and processes, failure analysis, scanning electron microscope analysis, corrosion, welding high-temperature alloys.

- **Moller, George E.,** 2224 Chelsea Road, Palos Verdes, CA 90274; 213/316-0023 (business), 213/377-1348 (home).
  Specialties: Corrosion and metallurgical engineer particularly conversant with heavy industry—oil refining, gas and oil production, electric utilities; pulp and paper; marine; chemical plants; high-temperature process equipment and repair; flue gas scrubbers, heat exchangers, piping, pumps, vessels, tanks, fired heaters, boilers, compressors, filters, dryers, evaporators, valves.

- **Monroe, Robert E.,** 2231 Franklin Street, San Francisco, CA 94109; 415/441-1798.
  Specialties: Weldment fitness for service, material selection and application, failure analysis, improved fabrication techniques, laboratory facilities and operation, welding and nondestructive evaluation quality control, materials and fabrication aspects of ASME, AWS, ANSI, and API codes.

- **Phillips, Austin,** 596 Dryad Road, Santa Monica, CA 90402; 213/454-3749.
  Specialties: Electron fractography, electron microscopy, metallurgy, aerospace materials and processing.

- **Promisel, N.E. (Dr.),** 12519 Davan Drive, Silver Spring, MD 20904; 301/622-3426.
  Specialties: Aerospace materials and fabrication, corrosion, advanced materials, resource conservation, national and international planning and policy.

- **Queneau, Bernard R.,** 2434 Berkshire Drive, Pittsburgh, PA 15241; 412/835-3980.
  Specialties: Iron and steelmaking, process control, heat treatment of steels.

- **Ray, Robert L.,** 1330 Broadway, Suite 1044, Oakland, CA 94612;

415/835-4432 (business), 415/339-1987 (home).
Specialties: Welding, heat treating, brazing, failure analysis, microscopy, mechanical corrosion; testing steels, stainless steels, superalloys; nondestructive testing.

- **Rosenthal, Philip C.,** 960 Parkwood Drive, Dunedin, FL 33528; 813/733-4932.
  Specialties: Metallurgy of iron and steel, heat treatment, metal casting, metallurgical thermodynamics. Teaches broad spectrum of subjects.
- **Sherman, Russell G.,** 592 Dryad Road, Santa Monica, CA 90402; 213/641-1723 (business), 213/454-2989 (home).
  Specialties: Expert in the manufacture, metallurgy, and use of fasteners. Experience in manufacturing using nickel- and cobalt-based alloys, titanium, refractory metals, high-strength steels, and heavy metals.
- **Shults, James M.,** 5305 Westcreek Drive, Ft. Worth, TX 76133; 817/292-2759.
  Specialties: Forging and casting—production, design, and inspection requirements; high-strength steel—processing, heat treatment, and inspection requirements. Primarily related to aerospace usage and problem solving.
- **Snyder, Harold Jack (Dr.),** 833 Jewel Street, New Orleans, LA 70124; 504/283-8565 (business), 504/283-1009 (home).
  Specialties: Alloy development and failure analysis—land, sea, and air.
- **Thomas, R. David, Jr.,** R.D. Thomas & Co., Inc., 103 Avon Road, Narberth, PA 19072; 215/664-8195.
  Specialties: Welding processes, welding of low- and high-alloy steels, nickel-based alloys, hard surfacing, weldments for elevated-temperature service; small business management—financial planning, acquisition studies, labor relations.
- **Wulpi, Donald J.,** 3919 Hedwig Drive, Ft. Wayne, IN 46815; 219/485-4853.
  Specialties: Serve as expert witness in product litigation involving metal failure; teach failure analysis and prevention; practice materials engineering.
- **Zapffe, Carl A. (Dr.),** 6410 Murray Hill Road, Baltimore, MD 21212; 301/377-7294.
  Specialties: General metallurgy—production and fabrication of stainless steels, superalloys, and refractory metals; failure analysis of metals, glass, ceramics, and other materials of construction.